ADVANCE PRAISE FOR *RE*

MW00560212

"When it comes to your body, Katy Bowman ⎯⎯ ⎯⎯ best friend you've been looking for your whole life. Having Katy in your life is like having a superpower."

—**Kelly Starrett, DPT,** *New York Times* bestselling author
   of *Becoming a Supple Leopard, Deskbound,* and *Built to Move*

"This book contains all the ingredients we need to improve our alignment and make our daily movement more nutritive."

—**Mark Hyman, MD,** #1 *New York Times* bestselling author of *Young Forever,*
   The Pegan Diet, and *The Blood Sugar Solution*

"Let this book be your field guide to greater physical and psychological flexibility, and enjoy the peace of mind, physical health, creativity and connection that moving more has to offer you."

—**Diana Hill, PhD**, licensed clinical psychologist and author of *ACT Daily Journal:*
   Get Unstuck and Live Fully with Acceptance and Commitment Therapy

"*Rethink Your Position* is a stellar guide to moving and living well today. It altered how I live my life and I'm better for it."

—**Michael Easter**, author of *The Comfort Crisis*

"Katy's whole body of work is pointing toward an updated physical education for the coming challenges....The self-care contained in this book is not designed to make you more comfortable where you are, like a sauna. This self-care can literally change your 'position' in life to a deeper, more realistic feeling of comfort in the world."

—**Tom Myers**, author of *Anatomy Trains*

"In this important book, Katy Bowman helps us to reshape our everyday movement and find comfort and ease in our bodies. I consider it essential reading for all humans!"

—**Chris Kresser, M.S., L.Ac**, *New York Times* bestselling author of *Paleo Cure*

"A must read no matter what your age or stage of health or fitness."

—**Jill Miller,** author of *Body by Breath* and *The Roll Model,*
   Co-Founder of Tune Up Fitness Worldwide

"A sea change is occurring in exercise and rehabilitation. There is a movement to microdose your movement, with the understanding that each moment of the day provides an opportunity to make your health a little better. In this work, Katy provides a comprehensive map of how to use those moments—exactly how to do it. This is such an easy recommendation for everyone seeking health."

—**Dr. Eric Goodman, DC**, creator of Foundation Training

"*Rethink Your Position* shows how to add healthy movement into even the busiest, most deskbound life, using nothing more complicated than your body, the floor, and the nearest wall or chair. With an infectious enthusiasm…Bowman shows how easy it can be to access a healthier body, brain and life."

—**Caroline Williams**, author of *Move: How the New Science of Body Movement Can Set Your Mind Free*

"If you've ever wondered why your back, knee, shoulder, neck—you name it—hurts, or wished you had more time to exercise, or wanted to feel more at home in your body, your search is over. Katy Bowman helps you take better care of yourself by helping you understand how your body works and the simple adjustments it needs to be its best."

—**Kate Hanley,** author of *How to Be a Better Person* and host of the *How to Be a Better Person* podcast

"Bowman is peerless at both understanding and interpreting human habits as well as finding new ways for us all to mindfully move. Told with typical humor, intelligence and brio, its practical, no-nonsense approach to innovative movement will have you head-ramping in no time. I loved it."

—**Professor Vybarr Cregan-Reid**, author of *Primate Change: How the World We Made is Remaking Us*

"This book contains the foundational principles to transform your daily life activities into opportunities to move and feel better."

—**Aaron Alexander**, author of *The Align Method*

"It's a simple, yet transformative prescription for living longer, better."

—**Maddy Dychtwald**, author and co-founder of Age Wave

RETHINK YOUR POSITION

Reshape Your Exercise, Yoga, and Everyday Movement, One Part at a Time

Katy Bowman

PROPRIOMETRICS
PRESS

Printed in the United States of America.
First Edition, First Printing, 2023
ISBN-13: 9781943370238
Library of Congress Control Number: 2022951610

Propriometrics Press: propriometricspress.com
Author photo by Mahina Hawley

Editor: Penelope Jackson
Science Editor: Andrea Graves
Cover and Interior Design: Zsofi Koller, liltcreative.co
Proofreader: Kate Kennedy
Indexer: Michael Curran

Cover Illustrations: Net Vector/Shutterstock.com; Good Studio/Shutterstock.com; Robuart/Shutterstock.com; Irina Popova ST/Shutterstock.com; Flash Vector/Shutterstock.com; Burkova Ekaterina/Shutterstock.com; Mary Long/Shutterstock.com; Yindee/Shutterstock.com; Bosotochka/creativemarket.com; Rawpixel/creativemarket.com

The information in this book should not be used for diagnosis or treatment, or as a substitute for professional medical care. Please consult with your health care provider prior to attempting any treatment on yourself or another individual.

Publisher's Cataloging-in-Publication
(Provided by Cassidy Cataloguing Services, Inc.)
Names: Bowman, Katy, author.
Title: Rethink your position : reshape your exercise, yoga, and everyday movement, one part at a time / Katy Bowman.
Description: First edition. | [Sequim, Washington] : Propriometrics Press, [2023] | Includes bibliographical references and index.
Identifiers: ISBN: 978-1-943370-23-8 (print) | 978-1-943370-24-5 (ebook)
Subjects: LCSH: Human mechanics. | Posture. | Movement education. | Exercise. | Physical fitness. | BISAC: HEALTH & FITNESS / Exercise / General. | HEALTH & FITNESS / Exercise / Stretching. | SCIENCE / Life Sciences / Human Anatomy & Physiology.
Classification: LCC: QP303 .B69 2023 | DDC: 612.76--dc23

To Nicole S., who ran alongside me and changed my mind about movement.

ALSO BY KATY BOWMAN

Grow Wild

Dynamic Aging

Simple Steps to Foot Pain Relief

Movement Matters

Diastasis Recti

Don't Just Sit There

Whole Body Barefoot

Move Your DNA

Alignment Matters

CONTENTS

PREFACE

A long time ago (fifteen years), in a galaxy far, far away (actually, this galaxy right here), I started a "why the way you move matters" blog called *Katy Says*. Five years and hundreds of thousands of words into writing it, I collected all the articles about body alignment and movement and put them into a book called *Alignment Matters*. That book was a favorite of many readers because of its short essay style providing quick, easy lessons on alignment and movement, written so that both movement professionals and laypeople could learn more about their bodies. That book covered a lot of ground and was very long, too long to be translated for my international readers. In the last ten years, a lot of the material in it has been developed into standalone books on different topics.

I have known for a while that I needed to write a new, improved, updated primer on the importance of alignment. And I wanted it to reflect where I stand on alignment now, a decade older and wiser—and to include new insights and all the best pieces that I've written since *Alignment Matters* came out ten years ago.

And so I am hereby welcoming to my book family *Rethink Your Position*, a fresh and lighthearted collection of accessible lessons about how to position your body in a way that allows you to better move through the world.

Enjoy!

INTRODUCTION

E very day we make hundreds of choices about how to move
our bodies.

Will we walk, or will we drive? Will we sit, or will we stand? Will
we slouch or sit up tall? Will we keep our hands in our pockets or
crossed in front? Will we wear heels or flats? All day long we make
choices about the positions we place ourselves in, and how often we
vary our body position, whether we realize it or not.

While disabilities might immobilize us or parts of us, by and large
we have uncountable choices to make about how we move. The
problem is, we make most of those choices subconsciously, usually
choosing the move that's easiest in the moment, and we suffer long-
term consequences for not being more deliberate in our approach to
using our body.

I've written many books about movement. Books on how certain
positions can lead to body damage and pain and how other movements
can make the situation better. Books about getting kids the movement

their body needs as they grow, about moving as we age, and movement as a form of activism. *Rethink Your Position* is also about movement.

More specifically, I want to get one big idea across to you with this book: The way your body works and feels is all tied up with how you move your body all day long, but you have a lot more choice than you probably know when it comes to how you move your body all day long, which means *you have a lot more control over how you feel in your body and how your body works* than you currently realize.

Rethink Your Position is about the options we have when it comes to positioning our body as we follow along in an exercise class, walk from here to there, or pull a suitcase through the airport, and it's also about the idea that our bodies are *always* being moved, even when we're just sitting there. Using better form all day long is another way to move different body parts, which means you can move more of your body—you can move your body more—while you're "just sitting there."

In fact, because you're probably sitting more often than you are exercising, making over the alignment you use to perform everyday tasks and activities throughout the day can make you feel really good. An hour of daily exercise done on purpose is great for the body, but so are minutes of thoughtful movement done every hour, hour after hour. By learning to consider your body in all that it does, you'll find yourself getting way more movement than just exercise alone gives you.

As a biomechanist I study how forces and motion affect *living* matter, and specifically human bodies. I know that the way our bodies feel each day, and the things our bodies will be able to do as we age, relate directly to how we've been moving. Not only *how much* we've been moving but *how* our body lines up with gravity as we move. Sure, we're also impacted by the quality of food, rest, and love we all get, but our *mechanical* environment—i.e., how often our body is feeling and dealing with loads created by its own weight—is constant. Our cells are sensing and responding to the mechanical environment every second of our lives, which means our movement habits are *super-duper* impactful.

Bodies start to hurt when they aren't moved enough, but also because when they are moved, some parts aren't moving with ease. This then makes it harder to move enough, and our movements get more diminished, immobility and pain arises, and we think it's all inevitable.

It's not inevitable. Our ability to make small adjustments in the way we create body loads, to change the state of our tissue through movement, our ability to change our alignment, is an inexpensive, low-tech, and effective way to disrupt that cycle.

If "move more, it's good for you!" were the primary message in this book, you could stop reading it now, because I know you already know that.

Here's the part I think most folks don't know yet: **You have much more control than you realize**—over how you move, how it feels when you move, and therefore how you feel each day.

Rethink Your Position shows you how the controls that move your body work. Once you learn them, you can steer your body in the direction you'd like to go.

A NOTE ON PARTICULAR BODIES

Often in response to my work on movement, I get questions about "what if?" What if I'm in a wheelchair? What if I have a chronic illness? What if I have a joint replacement or fused vertebrae? What if I work too much to move more?

My perspective is biomechanical, and I speak to our broad needs for movement in bodies that have been saturated in the same or very similar sedentary cultures. The fact is that while individual capacities for movement might be unique, every body needs movement.

While we all need movement, and many of us need much more movement than we're getting, we aren't all starting from the same place. There are bodies that spend most of the day seated and don't get any exercise or physical activity, bodies that exercise purposefully every day but spend the rest of the time sitting, and bodies that get a lot of movement at work: many people have jobs that keep them physically active, multiple hours a day—standing for a large portion of the day, repeatedly bending over at the waist to harvest row upon row of vegetables, climbing in and out of vehicles, or traversing up and down buildings.

There are bodies with parts that have never been able to move fully or at all due to disability, bodies with parts that have stopped being able to move through a lack of use, and bodies with parts that have stopped being able to move through overuse. There are both active and inactive bodies that hurt. There are bodies in pain that would benefit from moving more and bodies in pain that would benefit from not continuing to move in the same way over and over again.

My message for every body with every constellation of activity levels and restrictions or disabilities is this: Use alignment tools to observe how you're moving your body, and then you can move it better, part by part, to the best of your particular ability.

A NOTE ON THE IMAGES

Over the last twenty years I've found simple visuals work best when it comes to explaining nuanced movement. I've sent clients home with stick figures of the moves they should focus on and whipped out crude line and box diagrams for my students to get the quick gist of what's moving and in which directions. Lovely images they are not, but beauty isn't their point. To be able to convey complex, multi-planar movements in a way that allows most folks to pick up the lesson is the point. Images aren't to scale, but they're accurate enough to see at a glance where the hinges are and how parts move. I learned this approach from what I think is one of the best anatomy books, *Clinical*

Anatomy Made Ridiculously Simple—a text beloved by university students the world over. Finally, simple (silly) line drawings that clearly show how things relate to each other, no beautiful body paintings masking where things are. I hope my drawings are as helpful to you as that book was to me.

CHAPTER ONE
YOUR HEAD AND NECK

" One reason we get shorter over time could simply be that our overall weakness makes it harder for us to hold our bodies upright. But more importantly, those actual mass losses in our spine can result from years of not carrying our own body parts well."

START HERE

There are many ways to get more parts of your body moving, and not all of them require you to stop what you're doing. You can move your body more, right now, while you're *just sitting there* reading this book, starting with your head.

You don't have to be a biomechanist to have noticed serious posture problems in today's culture. Just take a quick glance around any public place and check out the number of heads that are looking down. Not only looking down with the eyes, but with the head. Not only a little nod of the head, but a big forward curve in the neck and upper back. Not only a few people, but many—maybe even most—people.

Looking down for certain activities is nothing new. Reading, sewing, cooking, art—these are age-old activities humans have had to look down for. And there's nothing actually wrong with looking down. It's an important movement and part of why our spine is made of separate hinges instead of one long, straight bone. These spinal hinges (our vertebrae) curve into this shape quite easily, as if we were made for this movement. What is new and does pose a problem is the amount of time we now spend with our heads dropped forward, and how little we do any other movement of the head, upper back, and even upper body.

I can't in good conscience send you off into this book knowing that you might slump over as you read it. So, just as the magical fairy gifts the journeying dwarf an enchanted token to ensure they make it through the dark forest unscathed, I give you the Head Ramp* posture adjustment.

Start by touching the top of your head.

Without lifting your chin or your chest, reach the top of your head up toward the ceiling while sliding your ears back over your shoulders, as shown in the photo on the right.

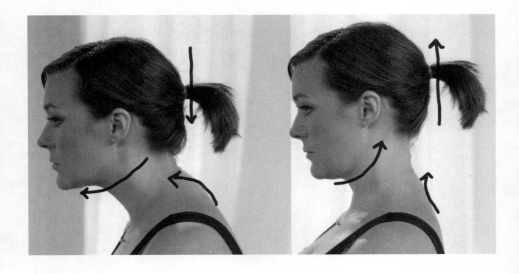

The head ramp movement is the *opposite* of the forward head shown on the left. It *undoes* a forward head by picking your head up away from the ground, moving it back over your shoulders and lengthening your neck. It is the *antidote* for a forward head and will keep you from being (posture-) poisoned on your reading adventure. Take it now, brave friend. It's time to begin.

*Does not work against gold thieves or dragons, although swords do, and if you've ever picked up a sword you know that good body alignment is required to aptly wield something so heavy.

THIS HEAD POSITION IS EASY TO SWALLOW

Think about your neck and all that it does. A neck is a great place to hang a necklace or drape with a turtleneck sweater, a neck turns the head, and a neck also contains tubes that move blood, air, and food.

The esophagus is the muscular tube also known as the "food pipe" that runs from your mouth to your stomach (and in the worst of times, from stomach to mouth, ewww).

Swallowing often becomes more difficult with age. There are a few reasons for this, but one has to do with the curves in our spine, which tend to change over time.

Imagine taking a mouthful of water then tilting your head way back and swallowing in that position. This pinches the throat-tube, which makes moving that mouthful down the throat harder and often uncomfortable, and it makes it easier to choke. Solid food is even harder to get down in tight spaces.

It's probably clear that when your head tips back your mouth and throat are no longer in good alignment. But it's less easy to see that when the upper back curves forward and your head faces straight ahead, this creates the same scenario: the head is tipped back relative to your spine, pinching the esophagus (left photo).

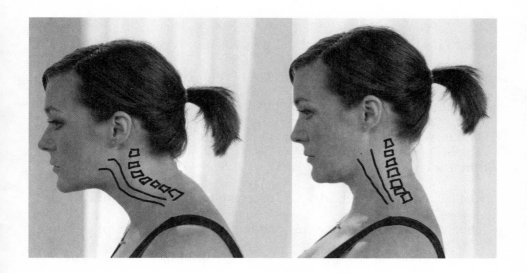

A classic recommendation for older folks in this situation is for them to tip their head forward—to look down—when swallowing. The idea is that if your upper spine and throat are curved forward, dropping your head forward to line up with your throat reopens the swallowing space, which reduces one of the mechanical hurdles to swallowing. And yet, another hurdle remains: the neck isn't aligned with gravity. Instead of the food dropping straight down, it has to move diagonally.

Temporarily dropping your head position for better swallowing alignment is a handy hack (an *anti-hack* hack), but why not work on improved swallowing with a head movement that opens *and* aligns the swallowing tubes at the same time? Meet the Head Ramp exercise, a single exercise that moves the head in three different planes all at once (right photo).

After you've nodded the head forward a bit to open the throat, work to lift the head toward the sky as you also try to move it toward the wall behind you...*without lifting the chin*. Not lifting the chin is important. If you do, you've just pinched the swallowing tube a bit.

And not only do you not lift the chin, you also try to sit up *without lifting the chest*.

Really give it your all, like you're trying to grow a long giraffe neck.

Then, to stretch tight muscles on the right and left sides of the neck, drop your right ear toward the right shoulder while continuing to think of lifting the head toward the ceiling. Then take the left ear to the left shoulder.

This "head ramping" movement will be subtle, especially at first, but it helps straighten your upper back and neck a little in the moment, and of course, more and more over time.

Good form is not only for golf swings. There are many places you can reduce your hacking, and eating time is one of them.

GETTING SHORTER WITH AGE

When my dad was eighty-eight, his upper spine was bothering him, so I accompanied him to see his osteopath for back pain. They began by weighing him and measuring his height.

My favorite part of this excursion was when they announced his weight, and he yelled, "Impossible! That's fifteen pounds too high! It's because I still have my shoes on," to which I yelled back (so he could hear me), "You wear fifteen-pound shoes? No wonder your back hurts."

My second-favorite part about this trip was talking about height change with him and the rest of the office.

Most of us will get a little shorter with age. This is usually chalked up to compression over time—either the discs between the vertebrae flatten a bit or the bone density of some vertebrae decreases and these weakened bones collapse, curling the spine forward. But there is another, barely mentioned reason for height loss too: geometry.

The compression-based reasons for height loss involve actual loss of the body's mass—disappearing disc fluid or bone. But you can also measure shorter *without* any change in mass. You can measure shorter or taller by changing your position.

Now, this book requires no math (you're welcome), but clearly, a height measurement is simply how high the top of your head is from the ground—it's not the length of your body. For example, I can take two S shapes that are the exact same length and change their height simply by deepening their curves. The body is like this too. Our knee, hip, pelvis, head, and spinal positions all affect our height a bit.

Here is a kindergarten-style model, complete with brads. I've taped the foot down, drawn a posterior reference line for head positioning, and then measured my model's height (Ht 1).

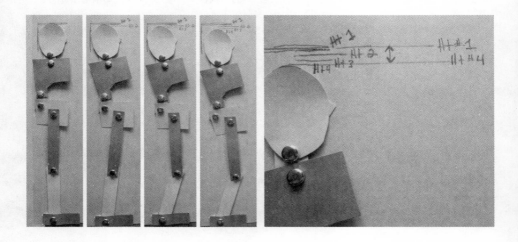

Then, as pictured left to right,

- I moved the pelvis forward (see page 128) and remeasured the height (Ht 2).
- I added a little knee flexion and remeasured the height (Ht 3).
- I added a pelvis tuck (posterior tilt) and remeasured (Ht 4).

What I couldn't show here (due to my low-tech kindergarten-style brads) was the further height decrease that happens when the upper spine and head move forward. But if you put that together with all the joint-position changes above, you'll lose the most height—no change in mass required.

So our position affects our height. So what? Well, one reason we get shorter over time could simply be that our overall weakness makes it harder for us to hold our bodies upright. But more importantly, those actual mass losses in our spine can result from years of not carrying our own body parts well. That means those traditional explanations of why we shrink aren't wrong, but they're incomplete, because mass loss—eroded bone density or dried-out discs—can be long-term responses to loads created by the same posture that makes us shorter in a minute. The mass-loss explanations don't recognize the impacts of using our body in a particular way, and why would we think to change our movement habits if they're never included as part of the problem?

My third-favorite part of this appointment was the troubleshooting between my dad and his doc. After examining his spine, the doctor asked him what lifestyle changes he had made around the time he noticed the pain. Dad noted he had just moved homes and, now that he mentioned it, his neck bugged him most when sitting and watching TV. After a few more questions the doc determined that the television had been placed too high in the new house. The effect was akin to lifting one's head to look through bifocals; Dad was wrenching his neck into a new position four to five hours each day.

The "medicine" he left with was movement: adjust the chair or television to create a new and better position that didn't deepen the curves of his upper back and neck.

YOUR PILLOW IS AN ORTHOTIC

My usual answer to "Hey, Katy, what's the best way to sleep?" is "As much as possible." But then I'll typically follow it up with something like this:

"There is no ideal sleep position in the same way there is no ideal 'all day' position. Humans have been sleeping on constantly varying natural surfaces, curling or opening their body for heat regulation, for eons, just like any other animal."

That being said, it is very uncomfortable to sleep on the ground when your body isn't used to it. Which brings me to the term "used to it."

Body training is something we tend to associate with fitness or athletics. Say you run a few miles every day, slowly increasing your mileage until your body is used to running five miles, then ten. In this case, the term "used to" means you've given your tissues and physiology adequate time to adapt. Saying "I'm used to it" really means "I've trained for this."

I've been playing with my sleeping environment over the last ten years to figure out the "sleep moves" that leave my body feeling best. Ten years ago, four days of sleeping on the ground in a tent or even on a carpeted floor would have left me with a neck spasm that would last a couple weeks. Why? Because I wasn't used to it. For thirty-five years I'd trained my body to sleep on a bed with a pillow.

We tend to think of human movement only in terms of exercising, but every movement or position you adopt feeds into your body's "adapt to this" file. Your tissues respond accordingly, whether the environment is "sleeping on this type of mattress for the last seven years" or "running seven miles every day." The process is the same.

The reason sleeping without your mattress and pillow can hurt is the same reason it hurts to run for miles beyond your regular running distance. You're placing greater loads on your body than it's used to.

Most of us feel very uncomfortable sleeping without a mattress or pillow because we were issued these comforts (read: these loads) quite young. We are given cushy pillows in the same way we are given cushy shoes. We all start using these items because everyone else does, and then we get used to them. The cells of our body adapt to shoes and pillows, and they fail to develop the strength and mobility needed to minimize and withstand the pressure of our own bodyweight pressing into the earth.

Personally, I realized that spending long hours every night in the repetitive neck position created by my pillow was the reason I was waking up headachy. It took me about a year to go completely pillow-free because I reduced my pillow height gradually. I went from something big and fluffy to something medium and fluffy. I progressed from a less fluffy pillow to a towel to a wadded T-shirt, and eventually to nothing but the sleeping surface and my arms to support my head as needed. If I woke up sore in the neck and shoulders after two or three nights spent at that height, I'd add back a little support.

I took a slow and stepwise approach, just as I did when training for my first half marathon. By progressing a little bit at a time, gradually loading the tissues in my upper body, I was participating in a "sleeping moves" workout.

Getting rid of your pillow is similar to getting rid of all your shoes and walking around barefoot all the time. Yes, you could throw out all your pillows and shoes immediately, but by doing so you might load underused tissues to the point of damage (as demonstrated by camping-neck or -back).

If you think your sleeping form might be negatively impacting how your body feels, work on exercises that help relax your neck and shoulder muscles during the day. This will increase their mobility so they cope better with a gradual decrease in external support at night.

Over the last six years I've trained myself to be able to sleep without a pillow (and without a mattress, for that matter). I've upped my sleeping game. Hotel mattresses no longer threaten me. Camping no longer worries me. I am more resilient now and am better suited physically to a wider variety of environments, i.e., my adaptation to my pillow does not keep me at home or require I haul a pillow every-where I go. The sleep-tone of my body is a lot better than it used to be, and this skill set is not really different from any other reason I strive to move more. I also get to train eight hours a night. Or sometimes four. During one phase of my life I did a biathlon where I alternated sleeping reps with breastfeeding ones.

Right now your pillow and mattress likely serve the same purpose as an orthotic or "supportive" footwear: they support you in positions your body can't comfortably support on its own. Support devices are helpful in many scenarios, but they can eventually leave those supported parts weaker. Doing all our walking on flat and level terrain wearing stiff, conventionally shaped shoes leaves us with stiff, weak feet; surrounding ourselves with cushions our entire sleeping-life has made our bodies unable to tolerate the movements created when our body adjusts to meet the floor.

Finally, know that our bodies are different day to day. Even years into my sleeping-moves journey, there are times when I'm injured or just physically spent for some other reason, and I seek more cushion or support. The point is to learn to adapt physically to different scenarios (which is one measure of physical fitness) as opposed to tolerating only a narrow range of physical experiences. Goodnight, and good luck.

CHICKEN HEAD

I got some chickens in the hope of being less of a biomechanist and more of an egg farmer. The first thing I did was spend a week studying how chickens move.

What's cool about chickens is that when you hold their bodies and gently move them—right and left, up and down, or in a small circle—their heads stay in place the entire time. Chicken necks are so flexible that chickens can keep their head still, like a gyroscope, while their body moves around it.

Their neck mobility is for good reason; they depend on it for their survival. In one way, we depend on neck mobility to be able to turn the head to look one place to another. A more expanded way of thinking about the function of neck mobility it is to think like chickens, who often need to keep their eyes fixed to something (like an eagle or a coyote) while their body moves toward safety in a different direction.

Although we rarely need to move away from predators, the ability to keep our eyes on something while we move our body in a different direction is practical for safe moving. Imagine moving through a busy city park, where there are cars, Frisbees, running dogs, and cyclists. If your whole body has to turn when your head does, it becomes difficult to walk in a straight line and you become vulnerable to tripping and falling. Even if you're not going to take your neck mobility out into the wild, necks just feel better when they're freed up from the shoulders.

Keep your ribcage and shoulders still while you move your head in these chicken-head exercises:

The Walking Chicken Head Thrust: A slow, forward-back motion of the head (think: pecking at the ground motion).

The Curious Chicken: You might think only dogs drop their ear toward the shoulder, but I have seen my chickens tilt their head to one side when they're wondering about something.

The Belly-Dancing Chicken: Sliding your head to the right and left without tipping it. And a real challenge: Keep your head ramped while you attempt the right-left motion! (Can't picture this move? Search "how to head slide" online and you'll find tons of tutorials.)

The "Is that a Coyote?" Head Turner: Practice turning your head to the right and left, ramping it first!

FIFTEEN PLACES TO HEAD RAMP

The Head Ramp exercise is a simple head movement that instantly decreases the load to the neck and upper back. This move fits into many scenarios and is one way to move more parts of your body all day long—whether you're tech-ing, farming, householding, or exercising.

Like all exercises, the Head Ramp is most effective when you're doing it. Below is a round-up of all the places you could think about better head positioning.

ON YOUR TECH

Phones, tablets, computers: Screens might be to blame for bad posture, but tech operates just as well with good alignment—we simply need to pass on the instructions for a better way to hold your head.

Game console: Part of a successful relationship to technology is understanding how it works. We are diving headfirst into our tech with scanty awareness. A few minutes discussing different body positions as a family can go a long way.

AROUND YOUR DWELLING

Car: Hours spent in a car provide good head-ramping time. Post a "reminder to ramp" note on the steering wheel or dashboard.

Kitchen: My bio leads with "biomechanist," but if we're going to list my work by hour, it should say "cook and dishwasher." Might as well stack that work with some upper-back and head movement!

Bathroom mirror: Add some Head Ramp when you're tending to your teeth or styling your hair, and use the mirror to check your form.

Music stand: Different instruments call for different head positions, but the idea here is to be aware that you might be able to hold your upper back and head better than you are now.

Workbench: Goggles on, head ramped!

Wheelchair: There's often a variety of ways to sit. As you're able, adjust your head and ribcage for better wheelchair alignment and put a visual reminder in an easy place to catch. If you're pushing a wheelchair, place it where you can be reminded to not lead with your head!

Sewing machine: Making can be hard on the body. You might need to hunch over to thread that needle, but a lot of the time you can be head ramping too.

WHEN EXERCISING

Mat: Head ramping is portable! Obviously you can head ramp while standing, but you can also do it in other poses too. Try head ramping in a plank, quadruped, or push-up position.

Bicycle: Stiff neck when bike riding? Try head ramping! When you lean forward to rest your hands on the handlebars, it's natural to want to thrust the chin forward to bring the eyes level to the horizon, but there's a better way. Instead, think of lifting the head to face the horizon by extending the upper back more. This will help lessen the degree of hyperkyphosis and hyperlordosis often created while bike riding.

Walking: Walking a dog, pushing a stroller or wheelchair, or on a treadmill/stair machine/elliptical, you can make sure your bout of movement is good for all your parts! Upright movement, especially walking, is a great time to pay attention to what the upper back and head are doing.

QUESTING

General quest tasks: While wielding swords against gold-robbers and dragons. Obviously.

CHAPTER TWO

YOUR RIBCAGE

" The mechanics of breathing can quickly become complicated—there's quiet breathing, breathing while sleeping, breathing during different exercise types, breathing through your nose or mouth, and it goes on and on. But all breathing involves the same baseline mechanics, and it all benefits when our breathing parts move better."

ARE YOU A RIB THRUSTER?

Let's talk about the body posture known as "swaybacked," hyper-lordosis, or excessive anterior (forward) tilt—all words used to describe a lower back curve that's too deep.

An excessive curve at the lower back is often associated with excessive loading to the spine, degenerative changes in the lower back, and lower back pain. One way many people are told to fix their extra-deep low-back curve is by tucking the pelvis under, which does somewhat straighten the lumbar spine and thus reduce the lumbar curve. But is a tuck always the best course of action?

I say no, especially if the excessive curve isn't being created by the tilt of the pelvis. You see, the curve of the lower back is affected by the movement of both the pelvis and the ribcage.

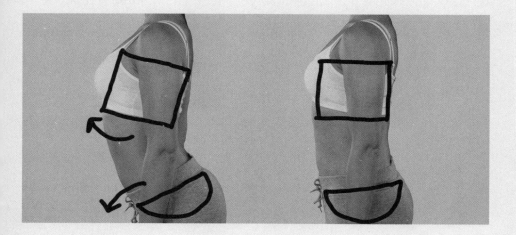

What about in cases where the extra deep curve in the lower back is created by movements of the upper body? In these cases, you can lessen the curve of the lower back by dropping the ribcage.

"Rib thrust" is the term I use to describe the motion of a ribcage that's lifted up. But it's more than a lift, because the ribcage connects to other parts. Lifting the front of your chest also lowers the back of your ribcage closer to the pelvis, which compresses the spine. It also shifts your entire ribcage forward slightly. That makes a rib thrust a *tri*planar movement that tilts and shifts a big part of your body in a way that's rarely helpful.

Moving the ribcage in this way also deepens the lumbar curve, so before you tuck your pelvis to "fix" too much curve in your lower back, make sure your curve isn't being made by a rib-thrusting habit. Tucking the pelvis under might provide some relief in the lower back because that action does decrease compression in the short-term. But its downside is that it changes how your feet, knees, hips, and pelvic floor are loaded by the body weight above them, in ways that can lead to injury in these parts when a tucked pelvis becomes an ingrained habit. *Mother tucker!*

USE THE WALL AS AN OBJECTIVE MEASURE

Rib thrusters generally have no idea they're thrusting their ribs, because it becomes an entirely unconscious habit. Use a wall to see more accurately where your ribcage sits relative to the rest of your body:

- Stand with your heels a few inches from the wall (more or less depending on the butt-space you need).
- Straighten your legs, make them vertical, and find a neutral pelvis (see page 124).
- Bring your upper body—shoulders and the back of your head—against the wall.
- Now check to see if you have a small space between your waist and the wall where your lower back naturally curves in.
- If your waistband is *entirely* on the wall, you are still tucking your pelvis. Tilt it forward (i.e., stick your butt out into the wall) until a space appears in the lowest part of your back. Once you have that space, bring your upper body back to the wall if it came forward.
- Now check to see if your middle back (the ribcage/bra strap/heart rate monitor area) is touching the wall.
- If your middle back is not touching the wall, move your ribcage toward the wall until it touches, allowing your head and shoulders to move forward if you need to.

In the image on the left, my pelvis and head are on the wall, but my ribcage isn't. Although my pelvis is neutral, I am rib-thrusting, and this creates a lumbar curve that is too deep for my body (#sway-back). In the right image, I have *the same pelvic position* (neutral)

as well as both my head and ribcage on the wall. You can see the decrease in my lumbar curve, no tucking of the pelvis required.

Maybe you can only get your ribcage back to the wall if your head comes away from the wall. This is common. It means you're experiencing too much curve and stiffness in the upper back. Ideally the head and upper spine should be able to move forward and back without moving the ribcage too, but that will take some work on the mobility of your upper (thoracic) spine.

This is another reason I'm not super hot about the "tuck the pelvis to fix the lower back curve" advice: it keeps people from seeing if their swayback is caused by too much upper-back curve. One way people attempt to compensate for a curved, stiff upper back (called hyperkyphosis) is to thrust their ribs—again, entirely subconsciously. But if you've identified extra stiffness in the upper spine, there are things you can do! To loosen it, focus on maintaining ribcage alignment (*ribs down!* is the simple cue) as you do your upper-body strengthening and stretching. This work will address the root of the issue.

Where to begin? Head ramping (see page 10)! Head ramping *without lifting your ribcage*! And chest-opening stretches, like Floor Angels (see page 43), *while keeping your ribcage from lifting*. Get it?

MOVE YOUR UPPER SPINE EVERYWHERE

I write a lot for work, so I'm on the computer a lot. One way I've found to stay productive is to regularly step back from my computer to do a minute or two of simple stretches. Our brains need the break, but so do our other parts. Have you ever found yourself breaking from the computer task at hand to quickly (and mindlessly) take a lap through your social media accounts?

If you're already breaking away from being productive, then congratulations, you've already found the time for an at-your-desk movement that will make you (and thus your work) ultimately better. So let's revisit one of my favorite office moves.

HOW TO DO A THORACIC (UPPER BACK) STRETCH

- Place your hands on the back of a chair, counter, desktop, or wall.
- With your hands in place, walk backward to lower your chest toward the floor.
- With your feet pelvis-width apart and pointing forward, back up your hips until they're behind your ankles. Bend your knees to take tight hamstrings out of the upper-body equation OR keep the legs straight to add a stretch to the hamstrings and calves.
- After you've done this a few times, rotate your elbows toward the ground for even more shoulder movement.

My favorite thing about this move is that you can control how much and where in your body you move. If you only want to move your upper back, don't worry about the arm position. If you want to concentrate the move in your shoulders, add the arm rotation. If you want to add your hamstrings and calf muscles, straighten one or both of your legs.

My second favorite thing about this move is that you can do it anywhere. Yes, do it after a long time on the computer, but also using the back of the couch after watching a full season of something on Netflix, on the kitchen counter waiting for your water to boil, on the car hood after driving too long, or on a fence post after you've dug fifteen holes.

The Thoracic Stretch is so versatile, so dynamic! Just like you are.

HOW TO MOVE YOUR
BREATHING PARTS BETTER

A Covid-19 infection or even the risk of one means that the past few years have been an ideal time to talk about breathing and the amount of oxygen you take in. If you know you're driving into a storm, you check your headlights and windshield wipers to make sure they're working as well as they can. Sometimes you don't know you're going to be driving into a storm (like me, EVERY TIME I visit New Mexico), which is why it's a good idea to maintain your car in general.

The breathing action is influenced by many things: environmental, psychological, biochemical, and biomechanical. Guess which category I'm going to write about now!

Breathing is a movement, and it's a movement affected—made possible, even—by the movement of other parts. The mechanics of breathing can quickly become complicated—there's quiet breathing, breathing while sleeping, breathing during different exercise types, breathing through your nose or mouth, and it goes on and on. But all breathing involves the same baseline mechanics, and it all benefits when our breathing parts move better.

ALL ABOUT BREATHING

Inhaling is getting the air outside your body to enter your body through your nose (or mouth) and travel down into your lungs. This happens when you lower the pressure within your thoracic cavity—that's the upper hunk of your body that contains your heart and lungs and is surrounded by your ribcage—so that air is *pulled* in.

How does air get pulled or drawn in? To lower the pressure in your thoracic cavity, you have to increase its volume. The thoracic cavity is like a room in your house: Your shoulders are a sort of ceiling, the ribcage and the muscles between them make the side walls, and the diaphragm muscle is the floor. In your thoracic "room," the walls, floor, and ceiling are flexible. When some of their muscles contract, the walls move outward and the space inside your thoracic cavity increases, and the more it increases, the more oxygen you pull in. You can lower the diaphragm to make more space at the bottom, rotate your ribs open to make more space at the sides, pull the ribcage away from the diaphragm to make room at the top, or you can do any combination of the three and get a bigger room all over. You want all of your walls to be pliable so you can maximize your thoracic-room volume to maximize each inhale when needed. And P.S. Just reverse all of this for the exhale—when supple parts move to shrink the thoracic room's volume, the lungs exhale, a movement skill that's beneficial not only for breathing (maximizing your exhale sets the stage for maximizing the next inhale) but also for a good, strong cough.

Our breathing parts can get stiff with disuse, just like other body parts mostly used to moving only a little bit. This is one of the reasons moving somewhat vigorously is so important. Oxygen-demanding movement is not only good for your heart and lungs, it's what keeps your breathing hinges OILED UP and free to swing as needed. You can't breathe well when the breathing parts are rusty.

WHAT ARE THE BREATHING PARTS?

A list of non-airway breathing "parts" would include a long list of all the bones, muscles, fascia, nerves, sensors, and chemicals involved, but some basic levers and pulleys include:

- the ribcage
- the individual ribs
- the vertebrae that interact with each set of ribs
- the muscles between each of the ribs (the intercostals)
- some muscles of the head and neck (scalene and sternomastoid)
- the diaphragm
- muscles of the abdomen
- muscles of the chest, shoulders, and arms
- muscles of the pelvic floor

Each of these parts helps or hinders the movement of other parts on this list, so when we think of expanding and shrinking our thoracic rooms enough to move the most oxygen, all these parts have to move.

Okay, now let's get your breathing parts moving more!

BREATHING EXERCISES

A simple way to move your breathing parts is to play with various breathing exercises. This is sort of like varying your sitting positions. It's something you're doing anyway but you vary it in order to move your parts differently. Breathing exercises can be as simple as trying to take ten deep breaths in a row or trying to breathe only through your nose, slowly. You can also find exercises online or remember any yoga breathing, or choir or theater breathing warmups. By paying attention to your breathing and by practicing breathing in different ways, you'll move differently.

However, your breathing will always be limited by any immobility of the parts that create a breath. That means sitting-and-breathing exercises can only take you so far.

Diaphragm/Abdominal Release

When you hold in your stomach, the contents of your abdomen get pressed up against the diaphragm, preventing it from lowering to its full range of motion and thus decreasing the total air volume you can pull in as you inhale. You're going to have to let your diaphragm down, and the only way you can do that is to let out any belly you're trying to hide in your abdominal cavity. Don't be shy. Bellies are beautiful, especially bellies that let you breathe!

Level 1: Seated or standing, allow your entire belly to relax, paying special attention to the sensation of your diaphragm releasing. You will probably feel (and see) your abdomen moving outward. Once you feel you've released your abdominals and diaphragm, try again—you might still be holding residual tension there. You will likely need to remind yourself to relax your diaphragm throughout the day.

Level 2: Relax your belly toward the floor while on your hands and knees (see pic below right). In this quadruped position, your abdominal contents are pulled more by gravity, and you can often let your diaphragm down a little farther.

P.S. When you are up and moving around, it's natural for your abdominal muscles to contract to support your movement. Abdominal tension created by your whole body movement is different from tension created by sucking your stomach in. Practice both so you learn to notice the difference. Once you're aware what sucking it in feels like, you can work to turn that specific movement off to allow your diaphragm to move even when up and moving around.

MOVES TO SEPARATE ARM MOVEMENT
FROM RIBCAGE MOVEMENTS

Floor Angels

This gentle action moves and lengthens your shoulders and chest parts.

Recline on a bolster or stacked pillows so your upper back is elevated but the bottom of your ribcage can still lower toward the floor. Reach

your arms out to the sides, trying to get the backs of your hands to the floor, keeping your elbows slightly bent. Once your chest can handle this stretch, slowly rotate the arms externally, so the elbows move up toward the ceiling and the thumbs point toward the floor. (Note: it's possible to turn your arms either way to make this movement, so make sure you're turning them backward.) Then, make a "snow angel" motion ten to twenty times, moving your arms along the floor (or however low you can get) toward your head then back down toward your hips.

Thoracic Stretch

Have you had your arms down by your sides most of your life? This stretch is a great way to start mixing up their movements.

Place your hands on the back of a chair, counter, desktop, or wall. Hands in place, walk backward to lower your chest toward the floor. With

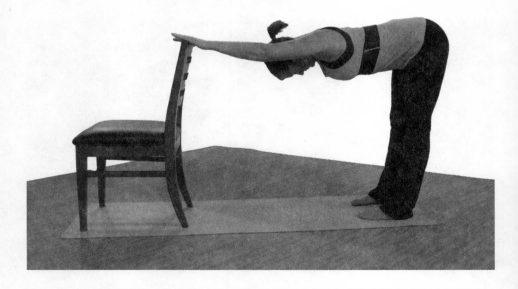

your feet pelvis-width apart and pointing forward, back up your hips until they're behind your ankles. Bend your knees to take tight hamstrings out of the upper-body equation OR keep the legs straight to add a stretch to the hamstrings and calves. After you've done this a few times, rotate your elbows toward the ground for even more shoulder movement.

Weighted Pullover

This moves your shoulders and makes the abdomen work to stabilize the entire ribcage (which is also good for breathing).

Lie on your back with your head and shoulders bolstered on a blanket as needed, legs extended on the floor. Grip the ends of a firewood-sized log (or bottle of water or any other weight) and squeeze your elbows toward each other (tight shoulders may make them want to poke out to the sides).

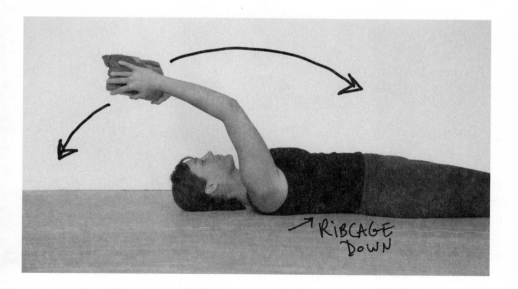

RIBCAGE DOWN

Keeping the bottom of the ribcage from lifting, lower the log overhead, only going as far as you can without letting the ribcage lift up toward the ceiling, then bring the log back up to the starting position. Repeat fifteen to twenty times.

Activate Your Intercostal Muscles

If you've been mostly belly breathing (dropping your diaphragm) and you have a tight upper body, the muscles between each of the ribs will have lost their ability to move. This is how you learn to strengthen them.

Set up: Seated or standing, firmly tie a stretchy resistance band or pair of tights around your torso just beneath your chest muscles or breasts (at the height of a bra strap or heart-rate monitor). Try to breathe through the nose for this if possible.

On the inhale: Try to expand the circumference of the ribcage and feel the resistance from the elastic pushing on your ribs. If you don't feel it at the end of your inhale, re-tie the band a bit tighter and try it again.

On the exhale: Try to *decrease* the circumference of your ribcage, pulling it inward, lessening the pressure from the elastic ring. Repeat, using each exhale to bring the ribs inward as much as you can.

You'll eventually feel muscles between the ribs and throughout the abdomen work on the exhale. You can use the elastic to teach you how to find and use these muscles, but once you've got the motor skill, you can do this exercise without the tactile assistance of the band.

A BIGGER MOVE

Arms have the anatomy to hold the weight of your body. Being able to hang from your arms is not only about fitness feats; the movements created by hanging affect your breathing parts too. If you're not ready to find a bar or branch and give it a go, start to work toward it simply by extending your arms overhead. You've already worked on a similar arm movement in the Thoracic Stretch and Weighted Pullover, but here's where you try it standing.

Doorway Reach

Every time you walk through a doorway, reach your arms up until you can touch the wall above it, then lower your ribcage back to neutral. Keeping the arms straight, step forward to increase the stretch, but really focus on dropping your ribs.

If you can't reach the top of a doorway, you can do a single-armed version of this exercise, one arm at a time, on the right and then left side of a doorway.

THE POSTURE

The way you carry your upper body as a whole impacts the leverage of many of your breathing parts. I'm putting "posture" last because although it's hugely important, I don't want you to wait until you've got your posture down to play with all the other movements above. Start moving all the parts above, then do them while aligning your ribcage (as outlined on pages 31–35).

KEEP BREATHING

Breathing is the most important movement of all. If we're doing it (and if you're reading this, you're doing it—WELL DONE!), we can breathe *well enough* when we're healthy, despite any resistances our body has. However, it's worth finding the sticky spots in our breathing so that if we're challenged, we can call on more breathing volume.

When we try to breathe in new ways, we ask new (and stiff) parts of our body to change shape, which can make breathing harder. That's because we feel the decrease in breath volume due to the stiffness of those parts.

Step one of *any* exercise is to keep breathing, so if you notice you're having trouble breathing, say, with your diaphragm released, or when you're trying to move your ribs, go back to the position where breathing is comfortable. Then try the same move over a smaller range. Maybe don't drop your ribcage all the way, just partway. Or maybe do the Diaphragm Release only partway for now and work on all the other moves for a while, then circle back to try them all again with your diaphragm a little more released.

THE WALL IS MORE THAN
A PERSONAL TRAINER

We can use very simple tools to objectively measure movement and body positioning. The wall is one of my favorite tools to use when teaching because it reveals so much. This wall-assisted move specifically helps with shoulder, rib, and waist mobility for better breathing.

SIDE BEND

Start in your personal version of standing up straight. Reach your arms overhead and, clasping your hands, bend your body to the side at the waist. Don't let your hips jut out to the side. Keep them anchored right above your feet.

Feel good? Feel any tight spots in the waist, ribs, or shoulders? Nice. Keep working on it.

Now stand against a wall. Set yourself up so your bottom (no tucking the pelvis!), your mid-back (that's the bra/bro strap area), and the back of your head all contact the wall. Note: If you have a lot of curve to your upper back, you won't be able to have both your head and the mid-back on the wall, so just get the mid-back on there for now.

Reach your arms overhead until your hands touch the wall above.

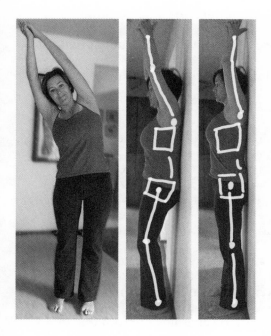

You might find that when you keep your body against a wall like this, you aren't able to get your arms up as far as you did in the first exercise. Maybe you're wondering, "How did my arms get up there before?" Many times we get our arms overhead by arching our lower back. That's because the lower back often moves to compensate for limited movement in shoulder joints. You might not realize tight shoulders can affect your lower back, or that you have tight shoulders at all. After all, they got overhead in that first exercise, and you might not have perceived how much your other body parts had to shift to get them there.

Now try the side bend, using the wall to keep your body parts in place.

How does being aligned for the exercise compare to when you *thought* you were aligned for the exercise?

The wall is really helpful here because it helps you realize which parts are moving and which aren't. We need to be able to catch which parts aren't moving, because those are the ones that most need our attention. Tools like floors, walls, straight edges, and alignment markers reveal sticky parts that otherwise fly sneakily under our radar.

There are practical reasons to discover when you unknowingly twist, thrust, or bend. After many years (and millions of little moves), these subtle movement habits can overload or strain areas and cause you discomfort and injury. Your awareness of how you move is key in changing such habits. Personally, the awareness a physical alignment practice brings me is also spiritual in nature. It helps me travel across the distance that separates *me* in my body from *me* in my mind.

As Thomas Merton said, "What can we gain by sailing to the moon if we are not able to cross the abyss that separates us from ourselves? This is the most important of all voyages of discovery, and without it, all the rest are not only useless, but disastrous."

CHAPTER THREE

YOUR SHOULDERS, ARMS, AND HANDS

> " What I'm trying to help you see is 1) how little your arms move in a day and 2) when they do move, how narrow a range they might move within. We could all use more total arm movement every day. More specifically, most of us really need more movements that get our arms overhead."

ARM SUBTLETIES

Put your arms out to the sides, making a T shape. Turn the palms up, then turn the palms down. Now notice how you are turning your palms up then down. This book is all about the subtleties of movement and why they are important. In this case, how you turn your palms up and down matters because you could be turning them up and down from the elbow (motions called *supination* and *pronation*, respectively) OR you could be turning your palms up and down at the shoulder (motions called *external rotation* and *internal rotation*, respectively).

When you weren't thinking about it, how did you turn your palms up and down? Now that you know there's two ways to do it, can you do both—turning your hands from your elbows and then turning them from the shoulders? Which is more difficult for you? Which needs more practice?

There are many situations like this in the body, where a motion like putting your hands palm down or up can happen in different ways, using different body parts. The important lesson here is, many times we depend on one way of moving our body (e.g., always turning your hands at the elbow) and the other one loses its ability (e.g., the rotator cuff muscles that can internally and externally rotate the arm bone get stiff, weak, and unable to turn your arm bones well). Once you learn what motions your joints are capable of, practice all of them. Just like your dentist will tell you that you only need to floss the teeth you want to keep, you only need to practice the movements you want to be able to do. WHICH IS ALL OF THEM.

WHEN EXTERNAL ROTATION IS OUT OF REACH

Our daily activities not only keep our body parts moving, they are great for checking in with *how* our body parts are moving.

Here's a movement test to try next time you need to reach behind you to tow a wheeled suitcase, pull a wagon, or take out the trash. When you reach backward to grab the handle, does your upper arm rotate internally, leaving your upward-facing palm to do the holding (image left), or do you rotate your upper arm externally so that your palm faces down before you grip and pull (image right)?

Think of the amount of time you spend—on the computer, driving in a car, sitting in a chair—with your arms out in front of you, palms down. Is it often? Is it most often? If so, the body tends to become more used to that movement than any other, and it can become less able to do other movements.

When your shoulders are tight, it's easier (in the moment) to reach back with the arm internally rotated and your palm facing up. Why? This way of moving avoids challenging shoulder parts that have become sticky from being stuck out in front so often. It gets those sticky parts "out of the way" of the reaching movement so they don't have to stretch.

Try reaching back with your palms facing forward (external rotation), then try reaching back with your palms facing backward (internal rotation). Notice which one is easier?

Both ways of reaching back get the job done, but they aren't equal. Internally rotating the shoulder every time you reach behind you is not great for the shoulder joint in the long term.

MOVEMENT MAKEOVER: If you find yourself always rotating inward during this movement, go a bit slower and try not letting your arm move this way. Grab something behind you, reaching with your thumb pointing out away from your hip until you get it. This will stretch the chest and shoulder more. It can be more work at first—but it's a stretch in the right direction.

FIVE STRETCHES TO SOOTHE PHONE-SCROLLING HANDS

Raise your hand if you're on your phone more than ever before.

Is your raised hand holding a phone? Then these stretches are for you.

Here are five moves that will get your hands moving more and differently from the phone death-grip, index-finger swipe your upper body has grown accustomed to, and bonus: *you have to put your phone down to do them.*

Stretch your thumbs. Make a loose fist with your right hand with the thumb pointing up. Grasp the thumb as low as you can with your left hand and move it like it's an old-fashioned Atari joystick, slowly moving it toward you and side-to-side at varying angles. "PEW PEW" noises not required. #80skid

 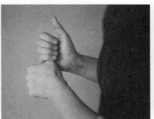 

Stretch your wrists. Keeping your shoulders down and relaxed, touch the backs of your hands together including the thumbs, then bring them down to waist level (image next page). Hold there or move them slowly up and down in front of your torso, or right to left. Keep those thumbs touching!

Stretch your shoulders and upper back. Place your hands on a wall, step back to bring your hips behind you, then lower your chest toward the ground to give your upper body an overhaul! This is one of my all-time favorite exercises.

Stretch your nerves. That's right, nerves need to move through their ranges of motion too! Reach your hands out sideways from your shoulders, making a T with your

arms and a "STOP" motion with your hands. Spreading your fingers away from each other, slowly work your fingertips toward your head. Keep your middle fingers pointing up, thumbs forward, and elbows slightly bent toward the ground. Think of reaching the upper arm bones away from you as you work your fingers back toward your body's midline.

And finally, Head Ramping. Keeping your eyes on the horizon, and without lifting the chin or chest, lift and slide your head back toward the wall behind you. This easy adjustment immediately decompresses the vertebrae in your neck by lengthening it, stretches the small muscles in the head, neck, and upper back, and increases your height. Back your face away from that screen!

SPRING FORWARD, SWING FORWARD

Ahhhh, moving the clocks forward in the spring gives us that extra hour of light at the end of the day. What do you do with that extra bit of light? I walk.

When I'm out walking more, I watch other people walk more. Gait patterns are made of more than just leg motion. The way the arms move when walking says a lot about how the shoulders and spine move in general. There are gait patterns with arms that don't move at all, arms that swing right to left instead of front to back, and sometimes there's one arm doing most of the swinging work while the other one mostly hangs down.

WHY DO OUR ARMS SWING WHEN WE WALK?

The way your arms move when you walk matters because their motion helps stabilize your torso and spine. Unless you're carrying something, your arms work in conjunction with your legs while you walk and run. When the left leg moves back, muscles in the back of the right arm contract to move the arm back too, then the right leg and left arm, repeat, repeat, repeat. After each arm has moved behind you, it relaxes and swings forward (not much work) to get into place for the next cycle. That's what happens on level terrain; arms do something different when you're going uphill, downhill, really fast, and really slow.

When you're walking on flat and level ground, ideally your forward motion is powered by your stance leg pushing off behind you (vs. lifting a leg out in front of you and falling forward, but that's a different essay). Our legs are heavy, and the motion of the left leg moving back quickly could spin the entire body to the left. Thankfully our leg motion is naturally balanced by the opposite arm. When your left leg moves back quickly, so does the right arm, which counteracts the spin and helps keep you walking forward without twisting back and forth. This balancing arm motion is called *reciprocal arm swing*.

A backward-swinging arm is not only great for reducing overuse of the spine, it is a built-in *workout for the back of the upper arm.* Awesome! If you thought three sets of twelve tricep exercises with a five-pound weight were effective, just wait until you get your reciprocal arm swing on.

PRACTICE STRENGTHENING, MOBILIZING, AND COORDINATING WALKING ARMS

- Stand and let your arms relax down by your sides.
- Without twisting your hips or shoulders, raise your right arm behind you to see how high you can get it. Hold for three seconds, then let it drop and swing forward passively.
- Repeat with the left arm.

Do this twenty to thirty times, then do both arms at the same time in the same direction, then both arms at the same time in the opposite direction, remembering the key here is to push your arms back actively but let them relax on the way forward.

WHAT ABOUT PUMPING OUR ARMS OUT IN FRONT OF US WHILE WE WALK?

Because we move so little, when we do move, we want to make it harder, which is probably where holding arm weights while walking came from. Or what about bending and lifting the arms up in front as seen in the sport of racewalking? How do these affect the balancing mechanism now that reciprocal arm swing is altered? In short, if your arm isn't going to go back to balance the leg motion, muscles around the spine need to tense up to keep your torso from twisting. So if walking has become a pain in your neck (or back), rethink the position of your arms!

If you are walking with weights because you want to work the arms, especially the oft-neglected back of your upper arms, then Rapunzel, Rapunzel, let down your arms! If your lower back and waist hold lots of tension, let your arms relax and swing naturally while walking (it may be that your torso locks up to stabilize your leg backswing because your arms aren't doing their part).

Improving your arm swing is going to change everything about your daily walk. And although the clocks might spring forward, our arms shouldn't. Unless you're walking downhill, but that's a different kettle of fish.

HOW TO CREATE A HANGING SPACE

Quick question: Where are your arms right now? Are they down by your sides? How many hours do you spend with your arms down by your sides? Even if your whole body exercises regularly, how do your arms move while doing so (walking, cycling, running, Zumba, etc.)? Do your arms move? How/in which direction are they moving, exactly?

That was more than one question, sorry.

What I'm trying to help you see is 1) how little your arms move in a day and 2) when they do move, how narrow a range they might move within. We could all use more total arm movement every day. More specifically, most of us really need more movements that get our arms overhead. In fact, why don't you take a break and reach your arms out to the side and then up overhead right now. I'll wait.

Once you've moved your arms differently for a bit, consider what you can do to remind yourself to do this more often, at work, at school, and at home.

The places where we spend most of our time don't encourage much arm movement. If you've already got the desire and intention to move more, then you need an environment that reminds you and facilitates the movements you've decided to do. Your next step is to create that space. If you want to move your arms more, I highly recommend a hanging station.

A hanging station is an area of your house or yard where hanging from a bar—or branch or whatever—is possible. Bonus points if you have to pass through this area multiple times per day; a hanging station in an out-of-the-way area will only move you when you specifically go there to move, whereas a pull-up bar in the bathroom or kitchen doorway is likely to prompt you to move your arms as you pass through the area.

Here's a list of things to buy or make that will help you add this movement to your home-life:

- Doorway pullup bar
- Rings (plastic or wood)
- Freestanding pullup bar
- Strong ladder, hung horizontally
- Sturdy branch
- Suspension trainers (like TRX)
- Finger strength/rock climber's training board

(Please note that setting up anything for you or your family to hang on takes certain knowledge. In many cases prefabricated items come with instructions, but when you're making things yourself, reach out to local builders or carpenters for assistance to ensure you've made or hung your items correctly.)

Reaching stations are also fantastic. You probably won't convert your entire home into a "primate habitat" (although, SURPRISE! It already is one) but in addition to devoting one area or apparatus to hang on, you can set up reaching stations at every doorway using a sophisticated piece of technology: Post-it Notes. They not only increase your reminders to move your arms more, they're also a great first step to hanging if you're not ready for a hanging station yet.

I love this Post-it Note action because these are essentially a form of TIME TRAVEL. My current self is sending clues to my future self. I have daily instructions left from the me in the past. That's exciting, right? You always wanted to star in a sci-fi novel (or is that just me?) and *now you can.*

Put signs in doorways you know you'll walk through and send a message *to your future self* to get those arms up and reach as you walk through.

Environmental modifications do not need to be expensive or time consuming; they just have to facilitate some sort of simple change.

(Note: Anyone with kids visiting or living in your primate habitat could consider putting some hanging tools within reach of kids' arms, too!)

HYPERMOBILITY

In college I learned that Hippocrates was said to have first referenced the idea of hypermobility in his writings about Scythians and how they couldn't effectively use their bows and arrows. One day I got curious about this and went to read what he wrote, but it wasn't clear to me whether he was saying that, so who knows for sure? What I do know (maybe) is that the terms used to describe a lot of our excessive joint motion—*hypermobility*, *joint laxity*, and *flexibility*—are often used interchangeably, even though they each mean something different.

The term *flexibility* typically refers to the ability of muscle to move through its range of motion, *mobility* is the ability for a joint to move through its range of motion, and *hypermobility* refers to a joint's ability to move beyond that normal range.

Every joint in your body has its own "normal" range of motion, while each person has a different ability to achieve it. This means some people will need to stretch to get to that range of motion, and others will need to learn to stabilize, i.e., not move so far. In both cases, more strength is helpful, but more on this later.

The extra mobility found in hypermobile joints is not always due to long or loose muscles, but to the laxity of that joint's ligaments. Said another way, being hypermobile doesn't automatically mean that your muscles are "too stretched"—in fact, in many cases it's the opposite. Often the joint's ligaments are just lax. So hypermobility doesn't just mean that your range of motion can be excessive; *the motions of the joint's parts* can be abnormal. In this case, *hyperlaxity* is an issue, in that your ligaments aren't holding the parts of your joints well, so now the *bones* are moving in a way they shouldn't.

IMPORTANT TIDBITS ABOUT HYPERMOBILITY, JOIN LAXITY, AND FLEXIBILITY

"Double joints" aren't an actual thing. But I bet you already knew that.

People are not hypermobile, *joints* are hypermobile. While there are conditions that affect the way all your connective tissue can tolerate a load, "hypermobility" is rarely a whole-body state. You might have some joints that have a greater range of motion, but probably not all of them. If you came over to my house, I would show you many joints in your "hypermobile" body that move less than the normal ranges.

Where do hypermobile joints come from? Nobody knows for sure. There are acute injuries and accumulated tissue change from years of using your body in a particular way. There are conditions like Ehlers-Danlos syndrome (EDS) that include some joint hyper-mobility. There is also a general class of symptoms labeled *joint hypermobility syndrome*.

Ligaments are the "seatbelt" of a joint. Like a seatbelt keeps your butt in your car seat, ligaments keep your parts connected and stop them from smashing through the proverbial "windshield" in a collision.

The ligaments are not the brakes of a joint, the muscles are. I like to call this out because just as good brake use in your car keeps you from needing to be saved by the seatbelt, the muscles around each joint keep you from giving their job to your ligaments.

The ligaments are not the brakes of a joint, the muscles are. See how important I think this point is? Just as good brake use in your car keeps you from needing the seatbelt, the muscles around each joint keep you from giving their job to your ligaments. Read this again and again until you understand what I'm saying here.

What am I saying here? I'm saying you need strong muscles to support the ligaments. I've known many people who've had ligaments removed and continued to move comfortably by using their muscles well. I mean, having both muscles *and* ligaments working well is ideal, but in this sedentary culture we keep trying to remedy painful ligaments and their issues without paying much attention to how unpracticed all the muscles are. We're essentially driving (read: moving) our bodies around with no brakes (read: muscles). Not only that, we don't know how to use the brakes (read: muscles), so we're depending on our seatbelts (read: ligaments) to keep our parts connected.

HOW TO MOVE WITH HYPERMOBILE (OR HYPERLAX) PARTS

If your joint stability is very poor, then it's likely your medical team will refer you to a physical therapist. But if you're cleared to go out and exercise in the world, here are some things to consider.

Folks who see themselves as hypermobile are often drawn to stretching and flexibility programs because these moves can be easy for them. The problem is, when entering into stretches, those with excessive ranges of motion often bypass tight areas and load those extra-mobile places again and again.

For example, one test for Hypermobility Syndrome is "Can you place your hands on the floor without bending the knees?" Now, many people can get their hands to the floor because their lower back is suuuuper mobile. At the very same time (in the very same body), many of these people have immobile hips and inflexible hamstrings. So, being hypermobile can also mean having really tight, immobile parts around those parts that can move a lot.

To move a hypermobile body well, approach each movement joint by joint, making sure you're positioned mindfully in every move you'd like to do. Sure, some moves might be off the table, but probably not as many as you thought. It just takes time to learn how to stabilize yourself so you move the extra-bendy parts less and the stiffer parts more.

PLANK MAKEOVER, STEP-BY-STEP

Maybe you already do planks and want to make sure you're putting the loads where you want them. Maybe you've always wanted to do a plank but aren't sure how to get there. In either case, here's how to check your elbows and shoulders to keep the work more in the muscles and less on the ligaments of the upper body.

Can your arms carry your weight when you're on your hands and knees? Start on your hands and knees, aka "quadruped position." Then put a slight bend in your elbows, which takes the weight off the arm bones and ligaments and gives it all to the muscles of your arms. It might sound super easy, but guess what? It's not. If you are someone with elbows that hyperextend, they're probably going to want to keep straightening, so keep watching them and replace that bend, over and over again. Spend a few minutes here (and each day going forward!) not letting them straighten. This will train your arm muscles to become stronger.

Once your quadruped feels stronger, progress to…

Plank-on-the-knees. A knee plank automatically requires your arms to work more than they did in quadruped, and keeping the elbows slightly bent in a knee plank is how to get them working even more. Do this a few minutes each day, letting your arm muscles adapt to carrying your weight. Progress when you can hold a plank-on-the-knees comfortably (*without hyperextending your elbows*, if that's an issue you have). Once you're stronger you can move back and forth between a slight elbow bend to your full elbow extension, but always with control and intention and not because you need to rest.

Now pair these moves with stable shoulder blades. Once you've learned how to stabilize your elbows, it's time to pay attention to movements between the shoulder blades. Your shoulder blades can squeeze together behind you or they can stay wide. When you're doing moves that require a strong torso and upper body (and that's a lot of them), work on keeping the shoulder blades wide. (It's difficult to see parts that are behind you, so try these next to a mirror where you can watch how your upper back is moving or ask a friend to help you when you're first learning how upper-back movement feels.)

Starting back on your hands and knees, widen your shoulder blades. Now, put that slight bend in the elbows. CAN YOU DO BOTH AT THE SAME TIME—keep your shoulder blades wide and hold your arms with that slight bend? If not, keep working towards that combination.

Once you can do this, walk your knees backwards to a plank-on-the-knees and manage both your elbows and shoulder blades at the same time. If you can't do both then walk your knees back partway and practice there.

Full plank. When you're ready for the next step (which could be anywhere between three minutes and three years) you can try a full plank on your toes. A recap: If stabilizing the area between your shoulder blades causes you to hyperextend the elbows, return to the level where you could manage both parts of your body well. Progress once elbows and shoulder blades are more stable.

Congratulations! You're learning how to hold your own body weight well. Once you can do that, you'll be ready to start carrying stuff that's not you, too.

RAISE YOUR HAND

Many of my good friends have had breast surgery, and the thing they like the least is the exercises they are supposed to do in the weeks and months afterward. What they like even less than those exercises that *they just don't want to do* is a bout of "frozen shoulder"—a painful condition that can arise after you immobilize your shoulder for too long.

Right after surgery, you need to be mindful of sutures. You can't just jump in and start moving, and your surgeon will tell you how long you need to keep movement to a minimum. "But after that," I nag my friends, "shoulder movement now is the oil for all your future shoulder movement." You've got to keep your shoulder moving (within whatever limits you are given by your surgeon as time progresses) because if you don't, you'll lose that shoulder's ability to move. Move it or lose it. Motion is lotion. That's why so many people in the movement field keep saying stuff like this: it's the truth.

I'm not going to give you any exercises here. You'll need to get those from your medical team based on the type of surgery you have. What I am here to do is explain why you're given those moves, what those moves are doing, and thus WHY YOU NEED TO DO THEM.

The first set of exercises you'll be given are picked to make sure your arm bone is moved relative to your shoulder blade in all the directions you'll want to move your arm bone in the future. They're most often gentle stretches or rhythmic movements that keep your shoulder joints from getting sticky. You'll be doing moves that get your arms reaching forward, over and over again. There will be moves that get your arms reaching behind you. You'll need to reach overhead. But not just overhead. You'll need to reach up and over to the right, and up and over to the left. The reason these moves have been selected is because they recreate the motions that make up the activities of daily living you'll want to do with your shoulders for the rest of your life.

Imagine trying to reach forward and up to put groceries away, or to grab someone's new baby to carry. To reach back to slide your arm into your favorite warm coat or scratch your own itchy back. We don't see people featuring these movements on social media accounts and they don't win awards at the Olympics, but they're the movements lives are made of. If there's something you can do to keep those movements, do it. I know therapeutic exercises seem like a nuisance, but trust me, you'll be happier if you have these movements available in the future.

In addition to the movements making your joints less sticky, movements in the arm are so very important to the lymphatic system. Your lymphatic system—part of your immune system that helps protect you from disease and infection—runs throughout the body. Clusters of lymph-processing stations (your lymph nodes) are located in areas of the body that are able to move a lot: at the armpits, groin, neck, and abdomen. Just like your arteries carry your blood, your lymphatic vessels carry lymph fluid. But whereas the blood moves through your arterial tubes with help from your heart, the lymphatic system has no big pump. Instead, the pumping action comes from the contracting and relaxing of the skeletal muscles the lymph tubes run through. When we stop moving an area of our body, the lymph in that area has difficulty moving out of that area. So the movement of your arm is key in keeping the lymphatic system moving, especially around the armpits and breasts so you don't get lymphedema. Get it? This is about more than your shoulder, although your shoulder movement, as we've already determined, is necessary too. SO JUST DO YOUR EXERCISES ALREADY.

Raise your hand if you'd like to be able to raise your hand. Exactly.

This advice—all the reasons you need to move your arms in all the ways—goes for everyone, breast surgery or not. We all want our shoulders to be mobile and our immune system to function well. While the process of joint stiffening might be accelerated in the case of surgery, it's always happening to all of us all of the time. Don't help the process along by staying still. Keep it moving, friends. Movement works!

BOOB BIOMECHANICS

In the U.S. there's a month dedicated to breast health (October) and a month dedicated to running (May) but there's no national running-breast month. If there was, I'd pitch an article about the biomechanics of breast movement and why runners (and vigorous walkers) should know about it.

Gait is a whole-booby phenomenon. Sorry, I meant whole-*body* phenomenon. There have been numerous studies on breast motion while walking and running, and it turns out breasts move in three different planes at once: up and down, side-to-side, and in and out. If you can't imagine the in-and-out motion, picture the way a punching balloon moves. Got it? You're welcome.

Running moves breasts more than walking does, and the amount of breast movement created by running increases as breasts get heavier. Movement of all body parts, even breasts—*especially breasts*—is necessary for their well-being. Breasts need to move: up and down, side to side, and yes, even like a punching balloon. Like all movements, though, there is always the question of load. What about heavy breasts? Long runs?

For people with exercise-induced breast pain from higher-impact movements like running and jumping, here are some things to consider.

SUPPORT

There is a net of ligaments within the breasts—Cooper's ligaments—that connect the breast tissue to muscles in the chest (pectoralis major). When you get the breasts moving for a bout of exercise, that motion pulls on the ligaments (and in the case of running, this pull is more like vigorous yanking), and this can create pain over time. This is where support can come in handy.

Look for sports bras from companies designing gear for active breasts, i.e., reducing movement in all three planes. This can also help with the other breast injury that arises with very active breasts—nipple chafing. Technology! You kids have it so easy! When I was a runner I just wore three bras stacked on top of each other, which held my breasts great…as they chafed huge cuts into the skin underneath both armpits while I ran to and from school in the snow, uphill both ways.

SMOOTH YOUR GAIT

Ever been to a conference or class reunion and been given a lanyard to put around your neck? And then walked down the hall, only to notice how it (or keys on a rope, or pendant on a necklace) bounces up and down, right to left, or in and out as you walk? WELL, YOU WILL NOW.

The lanyard that's *on* you, but is also able to move separately from the rest of you, reveals how your body is pushing it around. Breasts do the same thing. Breasts are part of us, but they're also sort of "on top" of the main body, freer to move than other parts. Now imagine if the lanyard was five or six pounds. Or twenty. Or forty. Now that extra side-to-side hip sashay with each step gets the breasts sashaying as well, only just a fraction of a second behind, so the timing gets thrown off and you've got this weight swinging up around your torso, which makes your entire body want to go in that direction a bit, and everything has to sort of lock down to minimize the effects. We end up walking differently than we would without the breast-response.

But we have breasts. So, to work on smoothing out our gait is to remove any *superfluous* movements—ones that aren't contributing to you being able to move forward—from the body so that your breasts (or lanyard) aren't being pushed around as much. But I do not recommend changing your gait to minimize the movement of your breasts. Researchers have found that many runners who are trying to minimize breast pain, movement, or both, squeeze their arms into their sides to hold the breasts in place. Or hold their torso with extra stiffness so their upper body doesn't move at all. That can minimize breast movement, but now you're messing with the efficiency of your running or walking gaits.

My take on things: If you want to pursue higher load activities and you find breast movement to be uncomfortable, wear a sports bra. Let the gear do the work during training sessions so the rest of your body can do the work it needs to do for the movement you've chosen.

STRENGTHEN YOUR CHEST

Do you need an athletic level of breast support all of the time? Probably not. Breasts need their movement too, and you come with anatomical parts that support your breasts. You can somewhat train breast ligaments by letting them feel and support the weight of your breasts on their own for some part of each day. And you can certainly train the muscles at the base of each breast: the pecs.

How much of your day is spent doing strong things with your arms that result in strong chest muscles lifting up those breast ligaments? How much of your day is spent sitting with your arms relaxed by your sides? Time to take breast health to the next step! One of the best ways to support bra-ed and unbra-ed breasts (and their health!) is to keep the chest muscles beneath them strong and flexible. Wide push-ups, hanging from a bar, chest-stretching. These movements offer better breast support from the inside.

CHAPTER FOUR
YOUR SPINE

"We are in a relationship with our body, and if 'move it or lose it' feels too negative, consider the same sentiment through a different lens: Care for it to keep it. We need to care for our bodies with movement, and by doing so, we get to keep and move with our body parts for longer."

EVERYTHING MOVES THE SPINE

Any movement you do with your body involves your spine. The cord of the spine runs through the vertebral column and conducts the messages that control movements of all your parts. Then the movements you do with other parts of the body move the spinal column right back.

I once heard that you're only as old as your spine is flexible. I get it, but I think the connection is indirect. I think it's to do with the fact that movement greatly affects the difference between our chronological age

(how many years we've been alive) and our biological age (the state of our cells, based on multiple things, including genetics and lifestyle). Regular physical activity naturally reduces many of the physiological markers associated with aging on the cellular level, which is why exercise is often referred to as "anti-aging."

Because many movements of the arms and legs involve your spine, a flexible spine is a good indication that other parts have moved regularly, which means you've moved your whole body a lot and are reaping the benefits at the cellular level. Or maybe the connection exists because the more flexible your spine, the more playful, "kid-like" things you can do with your body—get up and down from and roll around on the floor, hop down from things easily, and throw and carry stuff, all without jarring your back. It's probably both.

To keep your back tissues flexible, you need to move your spine through its ranges of motion every day, ideally more than once. That can look like this:

- bend forward as you do to touch your toes (flexion, see page 184)
- arch backward, lifting your chest and your chin *without rib thrusting*! (extension, see page 36 and image on previous page)
- lean your torso to the right and then to the left (lateral flexion, see page 51)
- twist to the right and left (rotation, see page 102)

Which movements does your spine do regularly? If it mostly sits upright in a chair, then you'll likely need to add all spinal motions to your "movement diet." If your work involves lots of forward bending (hello, food producers!), then you'll want to supplement with the other five motions you're not getting. Maybe you're a salsa dancer and your lower spine gets lots of movement but you could use some forward bends and extensions to move your upper back. If you regularly twist in one direction (looking at you, everyone in the dental field), your homework is to practice twisting in the opposite direction, plus all the other spinal movements you're not getting. GET IT?

GIVE THE SPINE SOME STABILITY

Remember when I said that moving different body parts moves the spine around? Our bodies have become quite stiff. We have shoulder joints, for example, that should allow our arms to move without tugging on the spine, but stiff shoulders mean that when the arms go to move, the ribcage has to come along. They've become overly bound or "coupled" together, when ideally they can go their own separate ways when they prefer. So stiff shoulders mean every time you raise your arms overhead, the vertebrae in the middle spine get pushed forward. *Arm motion* creates *spine motion*.

Our hip joints are so fabulous—they allow our legs to swing beneath us, moving us around all over the place without tugging on the spine.

Except that our habits leave our hip muscles so tight the hip joints no longer allow the legs to move around fully, so when we try to move our legs, they're all bound to the pelvis. Instead of our legs gliding beneath our pelvis, they are sort of fused to it, and the lower back becomes our new leg hinge.

The problem here is our spines aren't built to stand in for our shoulders and hips. Hips and shoulders are big and come with lots of support muscle. In contrast, our vertebrae and their muscles are small, and they're ill-equipped to take on the leverage forced upon them by bound-up hips and shoulders. So an additional step to spine care is working on improving your shoulder and hip mobility to make sure tightness in these big parts doesn't excessively move a few vertebrae in the spine. Note: It's common to have a spine that's stiff and under-moved in some areas and over-moved to the point of joint laxity in others.

After you've moved your spine in all the directions, you'll also want to work on shoulder and chest stretches (page 43), and any moves that separate the motion of the leg from the pelvis (pages 111, 193, 206 have a few to get you started).

It's not just arms and legs that affect the way the spinal column moves. There are the bones and muscle groups that pull directly on it:

- the shoulders and the ribcage
- the abdominal muscles that wrap around the midsection

- the diaphragm and psoas inside the torso
- the glutes and the big muscles of the legs that tilt the pelvis (and thus the lower spine) forward and backward

Spine movement is also affected by a broader connective tissue, called fascia, and the nerves that run through all of the back's parts.

All these parts are in a relationship. They are directed to move by the parts inside the spine, and they also affect the way the spine moves, which affects how the tissues are cared for, and around and around again. We are in a relationship with our body, and if "move it or lose it" feels too negative, consider the same sentiment through a different lens: Care for it to keep it. We need to care for our bodies with movement, and by doing so, we get to keep and move with our body parts for longer.

Just about every movement you do to position your body better improves how the spine is loaded. Changing to lower and more flexible shoes and stretching your calves affect spinal loads. Using a variety of sitting positions, walking more, and walking with better joint range of motion affect spinal loads. But in turn, these things that improve spinal loads are more difficult to do if your spine doesn't move well. How do you make a change, then? Improving spinal loads and spinal move-ment is not a linear or fast process. Rather, it's a gentle and gradual back-and-forth of small stretches that allow small increases in move-ment and small changes in habit, then more and more over time. Each change affects the other changes, bit by bit by bit. We're playing the long game, friends, but the good news is you're playing it with your best mate: yourself.

If you're not interested in more days spent with your spine, how about just feeling better each day, however many more you have? Back pain, especially lower back pain, has become a big, expensive problem worldwide. It's not just expensive in terms of dollars spent on health-care or missed work time; it's also, and I'd argue more importantly, costly in terms of our attention and the range of experiences we'd like to have. Dealing with back pain (any pain!) is exhausting and can keep you from *all the other things* you'd rather be doing.

I understand it's hard to make time to feed your body the movement it needs. Time is often our biggest hurdle. We don't want to step away from more interesting or pressing elements of our lives for boring old spine care. BUT TRUST ME. KEEPING YOUR BACK MOVING AS WELL AS POSSIBLE IS WORTH DOING. The spine, after all, is the backbone to movement.

WHAT'S A NEUTRAL SPINE?

Human bodies can move in all sorts of ways, thanks to our great number of joints. Many of those joints are found in the spinal column.

If we had one straight and continuous backbone, we would move our torso as a single unit (like we do when we've pulled our chin in toward our neck and have to turn our entire body to see behind us because the neck doesn't want to move). Thankfully our spine is made up of a stack of smaller bones—called vertebrae—that each move a small amount.

It's because of these small bones that human bodies can bend forward to crouch down low, roll into a ball on the floor, twist our heads or waists to look behind us, and lean our shoulders over to the right or left. A flexible spine is key to performing daily activities comfortably—as are muscles that support the spine (and the spinal cord and nerves that run through those small bones) in many different positions.

The spine is organized with lighter, smaller vertebrae up top and larger, heavier vertebrae farther down toward the tailbone. Even when we "stand up straight," we're not straight; the vertebrae are stacked in a way that forms a series of curves.

The natural curves of the spine include a slight kyphotic curve (a gentle forward "hunch") to the upper back, with curves in the opposite direction—lordotic curves—at the neck and lower back.

A spine aligned in these natural curves is called a "neutral spine."

This curvy spinal shape serves a purpose: it helps the parts of the spine—the bones, discs, ligaments, tendons, and muscles—carry loads efficiently and with minimal damage.

It's not only moving couches, carrying kids, and using the squat rack that stress the spine—adult humans are heavy, and our spines carry quite a bit of weight when simply moving our own upright bodies around. Learning how to stabilize your spine with your curves "in neutral" (which, again, does not mean straight) in a variety of positions is essential for loading your vertebrae and intervertebral discs in a sustainable way.

This is where things get a bit trickier, because it's not just the vertebrae and their

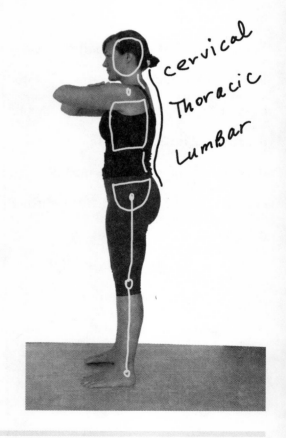

joints that affect the curves in your spine. The larger, heavier body parts that connect to your spine—your head, ribcage, and pelvis—change the curves of the spine based on how they're positioned.

Think about it. Tilting your head down to look at a device or lifting your chin to look through bifocal lenses changes the curve in your neck. Tucking your pelvis under to slouch in a chair curves your lower back. The vertebrae in your upper back connect to many ribs, so changing the position of your chest (and thus ribcage) brings all those vertebrae with it, increasing or decreasing spinal curves based on where it ends up.

But, in the same way I can't get my right eyebrow to move by itself, we have little ability to move the vertebrae in our spine individually. The curves are mostly created by what the head, ribcage, and pelvis are doing. Thus we take care of our spine by learning how to adjust and hold these heavier parts in a particular way, so *they* put the curves (and thus the bones and discs of the spine) where we need them to be.

FINDING A NEUTRAL SPINE WHILE STANDING

In minimal or fitted clothing, stand sideways in front of a full-length mirror in the way you might stand in line at the grocery store. Check out the position of your head, ribcage, and pelvis. Look beyond your flesh to find bony markers. Don't look at your hair, look at your skull. Don't look at your breasts, but feel around for the rib bones beneath them. Find the front of the pelvis or pokey hip bones. Do they line up with each other? Is the head in front of the chest or in front of the hips? Is the chest/ribcage lifted, or is it out in front of the pelvis? Is the pelvis out in front of both the head and chest?

Once you've assessed your "before" alignment,

1. **Shift your hips back** so your hip joints sit directly above your knees and ankle joints in a vertical line.

2. **Adjust your pelvis.** Your pelvis can tilt both forward and backward, but a neutral spine is created when the pelvis is neutral (neither tipped forward or backward).

   To find this neutral position, first locate the top, bony protrusions of the pelvis, called the anterior superior iliac spines (ASIS), and the bottom, front point of the pelvis, called the pubic symphysis (PS).

   Looking at your side view, stack the ASIS directly over the PS.

3. **Adjust your ribcage.** Your ribcage is somewhat cylindrical. Often, when we "stand up straight," we tip our shoulders back and move the bottom of the ribcage cylinder forward. Unfortunately this excessively increases the compression in the lower back, which is not great for the bones and discs in this area.

   If your ribcage is tipping back like the leaning tower of Pisa, tip the top of your ribcage forward to align the front of your ribcage so it is stacked over the front of your pelvis—which reduces the compression in your lower back at the same time. Be aware: If you are used to tipping the ribcage back, tipping it forward until it's straight will feel like you're slouching (the next step will help reduce that feeling).

4. **Adjust your head.** Spending lots of time looking down can mess with our spines. Whether we're looking at a screen, a book, a sewing machine, or a phone, our "dropped head" activities lower the chin toward the chest (which flexes the vertebrae in the neck) and move the head out in front of the body (which flexes the vertebrae in the upper back) *at the same time*. It seems like such a small thing, but this position we now call "tech neck" creates a flatter spine in the neck area and excessive rounding of the upper back, which can lead to diminished mobility and pain in these areas.

   To reset both the upper- and middle-spine curves to neutral, reach the top of your head toward the ceiling while also sliding your head back (don't lift your chin) as you bring your ears back toward your shoulders—all while keeping your ribcage in neutral.

   When you hold your ribcage in place, this upward-and-backward head motion tugs your spine up, and away from the ground, restoring the curves in your cervical and thoracic spine at the same time.

It is worth pointing out that we can use devices without rounding forward. While the content on a screen might beckon us to hunker our spines down around the electronic fire to be told a story, the buttons work in all the body positions, including the ones that are better for our backs.

It is also worth pointing out that our old posture habits aren't only deeply rutted in our brain; they are also rutted in the body. Don't fret when you find yourself constantly needing to remind yourself to move your head here or pelvis there. You can see these reminders in a positive light: Look how aware and attentive you have become when it comes to noticing and choosing where you're placing your body.

TAKING YOUR NEUTRAL SPINE FOR A SPIN

Once you've learned the basic adjustments to make, you can work on a neutral spine in many different positions and activities. A neutral spine is portable; you just keep the same relative position between the pelvis, ribcage, and head, and orient these to the planes of motion you're in.

A neutral spine while walking and running: This one is easy. To align your spine while traveling on foot, simply adjust your pelvis, ribcage, and head as you did while standing in place.

A neutral(ish) spine while cycling: Bikes, like other technologies, tend to move our body into a particular shape in order to use them. You've already learned you can ramp your head on a bike to reduce excessive curvature in the neck and upper back (page 26), and you can also reduce pelvic tucking a bit as comfort allows.

A neutral spine when on your hands and knees: Many exercises start with a quadruped, or "tabletop," position, and bringing a neutral spine to this hands-and-knees position can put you in a strong place to deal with loads that arise from a variety of exercises.

On your hands and knees in front of a mirror, practice tucking and untucking your pelvis. Watch how these tilting motions change your lower back curve—from a flat line to a deep bowl. Then adjust your pelvis until there's a small (not a deep) "bowl" at your lower back.

Keeping that small bowl shape, lift the bottom front of your ribcage up toward the ceiling until it is in line with the front of your pelvis. Holding your pelvis and ribcage in place, reach the top of your head away from your hips as you bring the back of your head toward the ceiling.

This lengthens your spine from head to pelvis and, again, restores the neutral curves of your spine.

A neutral spine while squatting or lifting: Squatting and lifting weights often require the body to lean forward. To find a neutral spine when you're doing moves like this, simply align your stacked pelvis, ribcage, and head to the torso-angle your exercise requires.

It's also important to note that "neutral spine" isn't a fixed position—there's a range.

Many exercises, especially lifting exercises, involve changing body positions throughout. In these cases, your spinal curves will also change; you're just working to minimize these changes by using the core musculature to stabilize your spine the best you can.

A neutral spine on your back: Lying on the floor, place a hand underneath the small of your back. Tuck and untuck your pelvis, noticing how your lower back moves toward the floor when your pelvis is tucked and how it arches away when your pelvis tilts forward.

Again, you're looking for only a small space under your lower back (i.e., a small amount of lumbar lordosis).

Note: If the muscles on the front of your thigh are tight, simply straightening your legs along the floor can tilt your pelvis forward a lot, creating an excessive lower back curve. In this case, you'll need to bend or bolster your knees for your pelvis (and thus spine) to be neutral.

Bring the bottom front of your ribcage down to line it up with the ASIS and PS on your pelvis. Now your ribcage and pelvis are aligned horizontally. Lastly, reach the top of your head away from your feet, which will lengthen your spine along the floor.

A STRONG, NEUTRAL SPINE REQUIRES A FLEXIBLE SPINE

A neutral spine is more than a shape, it's a state. When you keep all the parts of and around your spine supple and strong through movement, this "neutral" curve of the spine is the result. This is to say that a spine that finds neutral more easily is also moving through its other shapes with some regularity: rounding and extending as you get up and down from the floor, twisting and pulling as you turn, tucking into a ball as when sleeping, massaged by abdominal muscles contracting when you carry a heavy load, stretched long when hanging or reaching the arms way overhead. When you first learn about the idea that there's a better way to position your spine, it's easy to focus on trying to attain the position, but it's helpful to remember that the position we're seeking is the result of using the spine more dynamically. Thus, using your spine more dynamically (i.e., keeping it more flexible over all) goes onto the "healthy back care" to-do list, along with adjusting its position.

The ability to adjust your spinal curve depends on the mobility of the individual vertebrae. When we come to spinal mobility later in life, some parts of the spine can be stiff, making a complete "neutral spine" unattainable. In this case, make the adjustments you can, bolster your head or knees as needed, and spend time working on exercises and habit changes that specifically address the stiff spinal parts that are making a neutral spine less accessible to you. And, make the adjustments where you can without worrying about the others. There are so many different movements that benefit the spine, not being able to do some of them doesn't prevent you caring for your body.

To recap: Neutral spine is itself an effective position to use in a variety of situations, but not every situation. There are many movements that require the back to curl and extend and twist away from neutral, and it's doing these with more regularity that make a neutral spine easier to get into when you need it.

There's tremendous value to be found in the process of learning that your many parts are adjustable, and adjusting those parts you can.

Stable, strong spines that load the vertebrae and discs efficiently allow us to carry all our body parts in a more sustainable way, which allows us to move through life more easily, no matter what we're doing: standing at the sink washing dishes, bending over to lift a child from the floor, or walking through the grocery store to gather things for dinner. No matter what you're doing, you can carry your body better all the time.

A WAIST OF AN ABDOMINAL EXERCISE

I've taught exercise for what seems like a zillion years now, and I've seen people crunch, flex, strain, bulge, and jut, all to strengthen their abdominal muscles. Core strengtheners are great, but they're only half the equation. Your trunk needs to be flexible so it can be strong in multiple directions.

Our waist muscles have become tight. Many times they're so tight, they prevent us from twisting, which means the bones in the spine don't get to twist either. Moving your joints well is how you maintain them, so part of healthy back care is addressing stiffness in the muscles that wrap around and through the abdomen.

People tend to think of "abdominals" as mostly the long, up-and-down "washboard" abdominals (rectus abdominis) on the front of the body, but there are more muscles of the abdomen. There's the psoas muscle that runs through the abdomen, top to bottom. When this muscle is stiff, it can prevent the spine from twisting. There are oblique and transverse abdominal muscles that run right to left and diagonally on the sides of your abdomen, and which give you a waist. Sometimes our waist muscles (and thus the spine beneath them) are stiff. So stiff they, too, can keep the spine from rotating.

Don't just take my word for it. Start by checking how easily your torso tension allows you to twist.

Lie on your back and scoot your hips a tiny bit to the right, then bring the right knee in to stack over the hip joint (and keep it stacked there). Bring your knee and pelvis across your body to or toward the floor.

How much can *just your waist* twist to lower the knee? Can your upper body stay planted while you move your pelvis (image right)? This is a twist.

How much does *your entire spine* need to roll to lower the knee? When the upper body has to turn in line with your hips, this is a roll (image left).

SEE HOW THEY'RE DIFFERENT?

A twist moves your waist muscles so that they get better at moving; you don't want to roll during a twist if you're intending to create a change in your waist. Even if your pelvis doesn't go very far in a twist, at least you're starting to tug on your midsection, rather than always moving around sedentary areas.

Repeat on the other side to see how easily you move that way and where you feel like you're getting stuck. By doing this you'll learn to sense the difference between a Spinal Twist and a spinal roll and work on both.

Also, I tricked you—this exercise in measurement is also a set of exercises to practice.

Spinal rolls don't move the spine much, but they can still be part of a healthy spine program if you reach an arm behind you, away from the knee, where you'll feel if tight shoulder or chest muscles overly connect your arm to your torso (thus pulling on the spine as a whole).

Spinal Twists, even if they're small, help to maintain the rotational qualities of the vertebrae, especially if you keep the ribcage from thrusting and the back of the neck long (see: Head Ramp).

When we think of our midsection we tend to focus on how it looks, but how it *works* is much more vital. There are many important functions occurring in the midsection that are better when we move this area well: spinal stabilization, breathing, digestion, and local blood flow. Don't let middle movements go to waste, let them go to the *waist* instead.

HOW AND WHY TO PELVIC LIST

I once read a travel article that opened with something like, "Don't even think about traveling to Peru if you've got less than a month," and then it listed all the must-dos of a long trip to Peru. But the second part of the article started with, "Okay, so you only have three weeks? Here's what you have to do." The third part of the article was like, "Two weeks? Okay, do these," and so on, until it gave the *numero uno* tourist stop.

The question this article on Peruvian travel answers is, "How can I optimize my trip to Peru?" The author is saying that you should optimize it by allowing four weeks. But then they go on to offer how to optimize a shorter trip, and a shorter one, and a shorter one still.

I totally get this. I'm often asked, "What are all the movements we should be doing?" After I submit my answer describing how to move different parts all day long, I'll get a new parameter: "If everyone had an hour every day, which are the most important?" I have to scale the first request into a smaller amount of time, so I start crossing off items under the "must-do" category to try to optimize an hour's worth of movements. Then I'll get another request: "What can I do in a few minutes a day?" Now I need to pick the *numero uno* move that will make the trip worth it. Until now, I've almost always chosen the Calf Stretch.

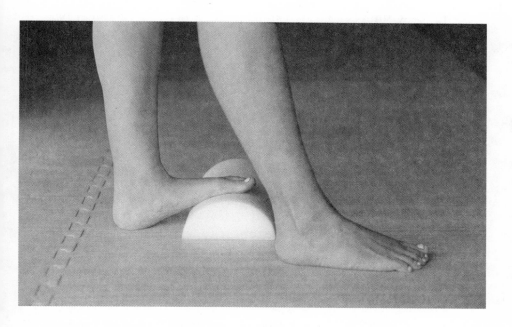

For years the Calf Stretch has been my number one exercise (see page 217 for instructions). For those thinking about moving differently and how that might change how they're feeling, the Calf Stretch is a simple, scalable move. (To scale your Calf Stretch: Change the height of what you're stepping on, how high you position your foot, or how far you step with your other leg.) It can change the loads created with every step; the mobility of your lower leg impacts your feet, ankles, knees, hips, spine, and, and, and. The ankle position created by a Calf Stretch is an element of every single step you take, whether over flat ground or going uphill. Even when we're pretty sedentary, we're usually still walking a handful of steps from point A to B to C, so the Calf Stretch really is effective for anyone who walks at all.

For those willing to tour two stops on their body, my number two exercise has long been the Pelvic List. The Pelvic List is a movement that both stretches and strengthens the muscles on the outside of the upper thighs. A strong pelvic listing motion helps you walk with balance, and it's how you get up and down stairs and hills without overloading your knees (see page 187). This exercise loads the bones of the hips in a way that keeps them strong (if you're thinking about bone density this is an excellent exercise to learn and put into every step) and offers more control of your whole body when you do single-legged exercises (if you wobble a bit, the large muscles of the legs can keep you stable). The muscles used during a Pelvic List run between the thigh bone and the sacrum. Strong listing muscles create a constant gentle, stabilizing tension on the sacrum at the sacroiliac joint, making this exercise an essential tool in your "care for my lower back" toolbox. Because the sacrum is also part of the pelvis, strong listing muscles also are essential for the pelvic floor, making the Pelvic List an essential tool in your "care for my pelvic floor" toolbox. Did I mention I'm talking about just one exercise?

Everyone benefits from a Pelvic List. Whether it's Goldeners who thought they'd lost their balance for good, athletes who couldn't get past their knee issues (the tendency to try to stabilize the entire body solely with muscles around the knees can prevent the hips from contributing), or new parents realizing they can access a heap of baby-lifting and -carrying help from their lower body, the Pelvic List is quickly transformative, over and over again.

All of the above has made me wonder if I shouldn't promote my number two exercise to number one. Just like you're not supposed to have a favorite kid (except for you, Mom! Sincerely, Your Favorite), I recognize every move offers its unique, valuable contribution to the whole.

BUT YOU WANT TO GO TO BODY-PERU AND YOU'RE ONLY GOING TO MAKE ONE STOP! Why let me decide where you go? *You're* ultimately choosing where you want to visit. I'm just giving you options. So here's a map to the Pelvic List.

As your tour guide, I'll let you know that "list" is a nautical term that refers to a boat rocking to either its right or left side, lowering one side of the boat. The Pelvic List is an exercise where one side of your pelvis lowers relative to the other side, so now you know why I named it that.

Start by standing with both legs straight, weight back in your heels. Push your right leg down into the ground. This helps create a contraction in the muscles on the outside of the right hip, causing the LEFT hip and leg to rise away from the ground. Slowly lower the foot back to the ground and repeat 10–12 times. Do this on the other side.

Once you have the lift, it's time to work on the lowering phase.

Stand with your right foot on a thick book or two or on a yoga block. Keeping both legs straight (no bending your knees), slowly lower

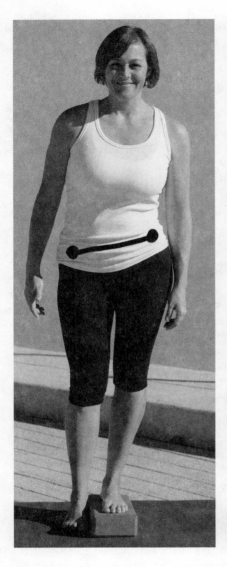
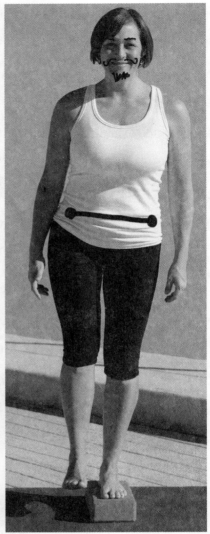

the left hip, trying not to touch down for a rest/balance check. Once

you've lowered, or listed, one side of your pelvis (image left), push the

standing foot down as you did before to bring the floating hip all the

way up (image right), then lower and lift it again until your standing

leg's hip muscles are fatigued. Switch legs and repeat.

It was a pleasure having you as my guest on the Pelvic List portion of your body tour. Don't forget to tip your guide!

DISCLAIMERS, ETC.

1. I don't know about my readers' individual bodies, so I work from the assumption that you can stand and have been cleared for basic movements like walking. I also assume you've been wearing shoes with a positive heel most of your life, have been sitting most hours of most days for most of your life, and don't walk many miles each day. These assumptions help me choose which exercise needs to be prioritized.

2. There's not really a hierarchy to these exercises; with hundreds of moves to choose from, the priorities depend on the individual body as well as an individual's goals and interests. Not having any of that information when I write for a wide audience, I have found a few simple daily moves that go really far. This is because **these exercises are designed to be impactful while you're doing them *and beyond*.** They're selected for their ability to change the loads you create all day long when you're not exercising. Meaning a few little moves can turn into hundreds of minutes of new movement each day. (And they're also easily slotted into things you're already doing, so you can actually do each exercise itself more.)

3. Further to that point, just like any vitamin shouldn't be taken in the absence of all other nutrients, if you only do one single move

ever, it can end up a) not helping much and b) potentially harming something. This is why cross-training is recommended for athletes, and it's why I suggest considering eventually traveling a greater area over your body—it's a magnificent place to explore! In other words, take your Pelvic List for a walk, introduce it to Calf Stretch, see if it hits it off with a Quad Stretch, and then hang out with them all on a monkey bar. Dig?

PLANES, TRAINS, AND A SORE BACK

It's not always clear which body part goes with which group. For example, the sacrum—the big triangle-shaped bone at the bottom of the long column of stacked vertebrae—is considered to be part of the spine, *butt* it's also one of the three bones that make up the pelvis.

While I'm not clear on how to categorize the sacrum, I do know long bouts of travel can lead to a pain in the joints around it, because of tension in the *piriformis*—a tiny but mighty muscle that runs between the sacrum and the thigh on each side of your body.

When the hips are fixed in a chair position for a long time, the piri-

formis can get stiff and create an ache in the sacral (low back? pelvis? buttocks?) area, and in the long term, stiffen and weaken the hip joint.

One move I always pack with me is what I call the Number 4 Stretch (so named because when you get into it, you kind of resemble the number four).

The simplest version of this exercise can be done right in your chair (see image previous page). Scoot forward so you're perched on the front of your seat, cross your ankle over your opposite knee, and tip the pelvis forward. This might already be a good stretch, but if you want more, lean forward. Note: **If you have a hip or knee replacement, you'll probably need to reduce or avoid this stretch.** If you want to avoid knee or hip issues, do the Number 4 stretch often! You don't even need a chair for this exercise (and P.S., you probably don't need any more chairs, ever), you can do this against a wall or you can do it standing and balancing and pretending to sit back into a chair.

P.P.S. Swap out any actual chairs with pretend chairs; they come in every shape and size!

You can whip out the Number 4 just like you whip out your neck pillow. You can't really do it while driving, but if you're the car passenger you can, and it's great to do while flying. Again, not if you're flying the plane, and probably not while you're in your airplane seat because there's too little room, but definitely if you're up at the bar and dance club in first class, or against a wall in the airport while waiting for your plane, or in the back of the plane where you can hang out with the flight attendants and get some extra pretzels because you look like one.

You can probably even do it right now, while you're reading this. Hint, hint.

PSOASANA

Savasana (also known as "corpse pose"), a yoga position where you simply lie flat on your back on the floor, is often labeled one of the most challenging yoga poses. Not because it's hard on the body, but because it's hard for the mind—to chill out that much, *to relax and just lie there* while continuing to stay engaged with ourselves or at least to not fall asleep.

A monkey mind can make it hard to stay focused throughout the pose, but it turns out many people indeed struggle with the physical part: savasana was included in a paper reviewing long-term injury reports following yoga classes, published in the journal *Mayo Clinic Proceedings*: "Soft Tissue and Bony Injuries Attributed to the Practice of Yoga: A Biomechanical Analysis and Implications for Management."

While savasana was in the bottom three of the list of twelve poses, it still had more reported injuries than headstands and splits, and it made it into this bit of research that reviewed the medical records of eighty-nine people reporting an injury following a yoga class. In fact, many people report anecdotally that lying on their back hurts. But why?

One function of our joints is to give our body access to various environments. In the case of lying down, joints throughout your body have to be able to articulate enough to give your many lying-down parts access to the floor. Simply put, in order to lie on the floor, you have to be as mobile as lying on the floor requires you be.

When the bulk of our hours are spent in the exercise of sitting in relative stillness, we adapt to this fetal-ish position. When we sleep at night, we do so on cushioned surfaces that accommodate this chair-shape, allowing even more hours in a similar position. (P.S. This is why I sleep on the floor. The firmer surface challenges the "being on the ground" mobilities of not just my joints, but of all my flesh.) The result of all this repetitive positioning? **Many of us are not actually lying flat when lying down on our backs, because we're not flexible enough to.**

Tight hips, legs, and upper spine (i.e., the physical adaptations to decades of chair sitting, among other habits I've gone on and on about) can make our lying down less like the image on the left and more like the image on the right.

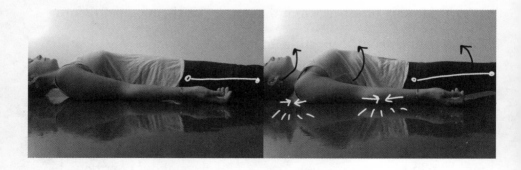

SOME MECHANICS

In general, lying on your back requires each of the vertebrae of your upper back to extend a small amount to allow your head and shoulders to reach the floor, and it requires the hip joints to be able to extend to 180° to stretch the legs out flat.

When all the lying-down movements can't happen fully, we can often still reach the floor in *some* areas, even if not all. The black arrows show some of the subtle movements that might be going on. If a tight upper spine can't relax down, the heavy head still drops, overextending the neck as it does so (note the chin lift), and the entire ribcage can rock backwards, lifting the front bottom of the ribs up in the air (note the ribcage rotation). Both of these movements can create compression in the spine, making this position painful.

Hips that don't extend to allow legs to meet the floor might reveal themselves by lifting the thighs away from the ground, even though the feet are down (see the arrow on the thighs). Buuuut, your legs are heavy. Even if your hips won't open, gravity keeps working to pull the thighbones down toward the floor. When overly connected legs do lower, they take the pelvis and/or the ribcage with them, which can further arch the lower back and compress areas of the spine. **This is one of the reasons folks feel better bolstered underneath their knees when on their back for a long period of time (while sleeping**

or during a massage)—it keeps the legs, elevated via tense parts, from lowering and pulling uncomfortably on other parts.

That's why something as simple as "lying down flat" can create excessive flexion and extension (and I'd also say vertebral shear) of the spine in some bodies.

You may have seen these displacements in yourself or, as a movement teacher, in your students. While some parts of the body can access the ground, chins, ribcages, thighbones, and vertebrae will displace in order to do so. Lying flat on command is not possible for many people; body parts well-adapted to a lack of movement take time to deal with. So maybe relaxing *is* hard for most of us, and not just in the sense that our minds don't want to chill out—our bodies aren't able, structurally, to quickly give way either (and surely there's a relationship between the two, right?).

An excellent way to make lying on your back easier in the moment is to add supports (pillows and blankets) beneath any body parts that aren't able to come down to the floor so they don't pull other parts in a way that hurts the spine. But this doesn't really address the issue of tight parts, so another to-do is adding an exercise designed to improve your ability to lie flat: the Psoas Release (see page 147). Let's call it psoasana. Or, to save time and better meet your particular needs, you can swap out the lying down/savasana part of a yoga or other exercise class with the Psoas Release. (Just one possible swap—maybe your psoas major parts move well, but other hip flexors, like the iliacus or rectus femoris muscles don't. Perhaps it's the mobility of the upper spine that is keeping you from accessing the floor. In these cases you could pick another move that gives you what your body needs most.)

Making a swap is sort of like bolstering, but it's also different. Bolstering is often done to make a move more accessible, but it doesn't always change *why* a move is not accessible. It doesn't always get you to the point where you can do the move with more movement. The approach here is to do both—make a move more accessible and increase your body's movement—at the same time. Create an environment that helps you find and change *why* savasana isn't working for your body.

This also takes care of another issue long-time readers often ask about: how to fit the corrective moves into classes of different modalities. If folks are coming for yoga (vs. restoring larger movements to fit into life), try trading out savasana for "psoasana"** if you see the need in your body or a student's body. Psoasana is not necessarily an easier move—it's simply a move that deals specifically with helping students become aware of any "chair baggage" they might have brought to class, giving them space to experience and address those loads in a safe way, while also working on downregulation and other benefits found in savasana.

Is savasana (or just lying on your back) "hard"?

There's debate, mostly in my mind, about labeling lying down on the ground as a dangerous or advanced move. On one hand, it is very important, so important I'm writing about it specifically, to recognize that lying down is hard in the sense that it's currently inaccessible to many peeps. This issue needs to be dealt with on the individual and movement-professional level in the form of more complete instruction.

On the other hand, it's essential we see the relationship between the environments and movement habits we promote as a society, and their effects. This movement is not hard by definition—it could be thought of as a very basic, non-intense, non-injurious movement humans have been doing for a very long time. But now, because of the way we haven't moved, this simple resting pose has become hard—too difficult to do comfortably—for many bodies.

We need to understand the origins of our movement difficulties. If we instead perceive a movement as intrinsically unsuitable, we risk eliminating even more movements from our culture rather than eliminating the habits that result in their difficulty. **Eliminating difficult movements is one way sedentary cultures can become more sedentary over time.**

As a teacher of movement, I believe that more complete movement instruction involves more than adding bolsters or swapping moves; eventually we need to be able to teach the context and non-exercise portions of human movement too. For example, the Psoas Release (or *psoasana*) can be used as an exercise to help make other exercises possible, but it's critical we, as a group, start thinking outside the exercise box. There are certain age-old human movements, like lying on one's back on a hard surface, that in our current society are becoming less prevalent, and we are therefore becoming less able to do them.

Rather than exercising solely for fitness, you can exercise using moves that can help you restore some of those movements *and the experiences they facilitate*. The Psoas Release exercise can make savasana more doable, and it can also help you lie on the bed or floor more comfortably and successfully. It can help with a good night's sleep. It can help you take a rest beneath a tree or lie down and look up at the stars. It can help you get under your car to change your oil, and any other fun supine activity you can think of.

Put the Psoas Release anywhere you want. Do it as a standalone (or is it a liealone?) move, or use psoasana as an easy and restorative way to end any movement session. It's a small move with big benefits.

The authors note that a pose being correlated with an injury does not necessarily mean the pose caused the injury—it's tricky to name anything as a "cause"—but these were the moves during which individuals experienced their particular issue.

**There's a range of reasons why yoga teachers end their class with savasana; the substitution suggested above might not be appropriate in every class. I am only speaking here to the physical shapes of a yoga practice.*

CHAPTER FIVE

YOUR PELVIS (AND PELVIC FLOOR)

" The pelvis is the junction of many parts, including those seemingly far away (for example, did you know there are muscles connecting your pelvis to your shins? So movements of your knees and shin bones move your pelvic-floor muscles!). Your pelvis parts are affected by all sorts of whole-body movements—and not only obvious movements like running or jumping, but even more subtle things we don't even think of as moves, like breathing."

LANGUAGE OF THE PELVIS

Every field of study has its own jargon. The study of movement has it too, in the form of words that help clarify the location of parts and the direction they are moving. Words like *pronation* and *supination* or *flexion* and *extension* do a more accurate job of explaining specific motions than other directional words like "up" or "down" or "to the right or left."

Anatomyspeak also includes terms that describe the position of a limb relative to a joint, like *abducted* (when a limb moves away from the body's midline), *adducted* (when a limb moves toward the midline), and neutral. *Neutral* is shorthand for *anatomical neutral*—a skeletal position that allows that part (or set of parts) to move maximally in all directions.

A "neutral pelvis" refers to a position of the pelvis that maximizes the degrees of potential tilts and twists of the pelvis and its surrounding joints. So while neutral is measured as a fixed position, the value of this bony arrangement is maximized when you move.

Yet it's not always clear how to measure a neutral pelvis in a practical way. In the beginning…or, at some point fortyish years ago, it was commonly accepted that the human pelvis was neutral when two points in the front of the pelvis (the *anterior superior iliac spines*, or

ASIS) lined up in a horizontal plane with two points on the back of the pelvis (the *posterior superior iliac spines*, or PSIS) when viewed from the side. This alignment places the pubic symphysis slightly in front of the rest of the pelvis. Later, a neutral pelvis was defined in other texts as the ASIS lining up with the pubic symphysis in a vertical plane (left image).

Neutral   Posterior   Anterior

I like to use the latter, although many people still use the former. And really, who knows which is best? The body didn't come with a manual, and the difference between the two alignments is subtle and likely not very important.

My reason for choosing to use the ASIS and pubic symphysis as markers of a neutral pelvis is mainly practical. Unlike the ASIS and pubic symphysis, the PSIS is not easily seen or felt. It's often buried under flesh and doesn't jut out as much as the top of the pelvis and pubic symphysis do. Even clinically trained professionals can struggle to reliably identify these posterior prominences, so it's probably not the best tool for everyday people trying to check their own pelvic position.

There is also jargony anatomyspeak to describe the movement and position of the pelvis, but some of that exists for good reason. Consider *anterior pelvic tilt* (for when the pelvis tilts forward; right image) and *posterior pelvic tilt* (for when it tilts backward; center image). *Anterior* means toward the front of the body and *posterior* means toward the back. In both cases they refer to where the *top* of the pelvis is heading. This is vital to understand, because when the top of the pelvis tilts in one direction, the bottom of the pelvis goes the opposite way, so during a "forward" or anterior tilt, the base of the pelvis moves in the posterior direction. We need to keep in mind which part of the pelvis's movement we're describing when we're using anatomical terms.

This brings me to less jargony ways of talking about pelvic motion: tip and tuck. Because the pelvis is shaped sort of like a bowl, it's simple to say, "Imagine your pelvis is a bowl of soup. Tip your pelvis forward like you're pouring the soup out in front of you [anterior pelvic tilt], or tip your pelvis back as if you were pouring it behind you [posterior tilt]."

Another trick is to think about the tail of a dog. A dog's long tail makes the tuck of its pelvis—when it's nervous or submissive—much more obvious. Thus movement teachers will often instruct students to tuck the pelvis when they want a posterior tilt and to *untuck* (get your tail up and wagging!) when they want an anterior tilt.

To recap:

An anterior tilt of the pelvis: untucked or forward-tipped pelvis.

A posterior tilt of the pelvis: tucked or backward-tipped pelvis.

A neutral pelvis is in between.

A neutral pelvis is part of a neutral spine, and having both in neutral helps optimize loads for a variety of activities. You might have too much anterior pelvic tilt at rest, in which case you'll need to work on tucking more in order to get to neutral. Or you might have too much posterior tilt at rest, in which case you'll work on tilting forward more. Some exercises have pelvic tucking as one of the steps, and some need an untuck.

That's a wrap-up of the front-to-back pelvic motion jargon: tilting, tucking, and tipping. But there are other pelvic motions. When the right half of the pelvis is higher than the left, say because you've been standing in line for a long time and have pushed your right hip sideways to make standing easier, we'd say the now-higher right hip is *elevated* and the left, lower one is *depressed*. And P.S. sometimes this pelvic position is also called a tilt (confusing!) but it's clarified by calling it a *lateral* tilt (lateral means away from the body's midline, or sideways, in anatomyspeak).

When a pelvis is tilted laterally, it's not neutral. The right and left ASIS sit at the same height in a neutral pelvis.

A pelvis can also twist, as you Latin-style dancers will know. That twisting motion is called *rotation* in anatomyspeak, which is pretty straightforward. As you probably guessed, a neutral pelvis does not have one of its halves rotated out in front. If you were standing facing a wall, the ASIS on the right and left side would both be the same distance from the wall.

When it comes to rethinking the position of your pelvis, you can adjust it in multiple ways: the tilt front to back, the height of the right and left sides, and rotation of one side relative to the other.

Those aren't even all the ways you can adjust the pelvis, but they're enough *for now*.

MIND YOUR PELVIS

There's another way to adjust the position of the pelvis: by backing up the hips (and thus pelvis). This *translation* of the pelvis changes its position relative to the ground. The pelvis might also tilt along the way, but that tilt is a separate movement from the translation. So let's get into this adjustment.

When we think of "good posture" we often think "shoulders back," but when we *do* "good posture," we're often not pulling our shoulders back so much as lifting the ribcage and pushing the pelvis forward. Our "standing up straight" tends to include a lot of pelvis forward. This can seem like a subtle difference, I know, but from a mechanical perspective, these chronic positional habits alter the angles of your body—up to 30° in some places!—which correspondingly alters the way your body is loaded and works.

How's my posture in the picture to the right? It's tough to say if you're not sure where to look or how to evaluate. I'm definitely not slouching, but is this all that it takes?

When we use objective markers it's easier to understand physical loads. When I add a line through the ankle, knee, and hip joints to that same photo, you can see this posture is "standing up bent" instead of standing up straight, and how the relationship between my upper and lower body changes when I reposition my pelvis back over my hips to create a vertical leg.

 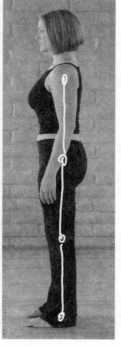

If you're evaluating your alignment or someone else's, you can hone your mechanical eye to look for the relationship between the ankle, knee, and hip joints. Then extend a line through those points to see if there's a forward (non-vertical) lean to the lower body. To adjust, just shift your hips back until those three points stack directly over each other. You'll need to be unshod or in completely flat shoes.

When the major axis of the body forms a plumb line, this "anatomical neutral" position maximizes the structural support of your bodyweight as well as your movement potential in all directions.

BACK UP FOR FEEL-GOOD FEET

Using that same mechanical eye, draw a vertical line from the hip joint to the floor. This helps show that where you wear your pelvis affects which parts of your feet feel the pressure of your weight. The forward pelvis in the left photo (below) places the burden on the front

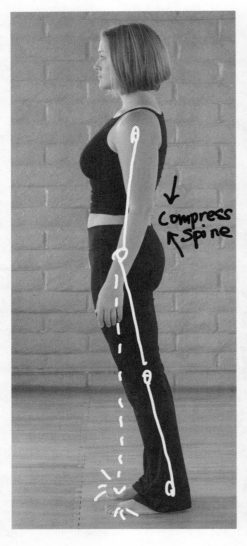

Compress Spine

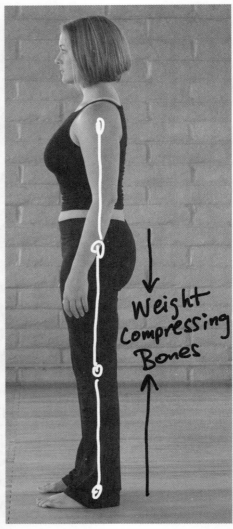

Weight Compressing Bones

of the foot where the smaller bones and muscles have to deal with it, and the backed-up pelvis in the right photo places it on the much bulkier and denser rear foot—freeing up those smaller foot parts and fascia for better balance, movement, and any healing they might need.

Backing up your pelvis is easy, fast, and free, but you might have already noticed that you can't do it while wearing shoes with an elevated heel. Shoes with a heel automatically force you to move the pelvis forward, so try backing up your hips while barefoot, bringing your weight far enough back to lift your toes off the ground. Now you have a sense of your weight being over your heels.

BACK UP FOR WEIGHT-BEARING HIPS

When we think of weight we usually think of how much we weigh (which seems right, right?), but weight is really how much of your mass is being pulled on by gravity in a vertical direction.

When my pelvis is not over my legs, the weight of the parts above my legs is not pressing on the bones of my legs. In this position gravity ends up making an archery-style "bow" of my body rather than compressing it. When I back up my hips to sit directly over my legs, my weight sits above the leg and pelvic bones, meaning my hips are now bearing my weight. My body mass is the same in both pictures, but because weight is a vertical force, the less vertical you are, the less you "weigh" from the standpoint of your legs, hips, and pelvis.

Who wants *weight-bearing hips?* We do. We need it for the density of these bones. But we barely ever carry the load of our body in a weight-bearing way. Sitting takes the load off our hip joints, making them bear little weight. And then, even when we do move or exercise, it's quite possible that our alignment means our hips still aren't carrying much weight. Our culture has major issues with bone loss. Why might that have come to pass, in a chair-sitting, raised-heel-wearing culture?

Your first stop on the osteogenic (bone-building) train can be *carrying your body weight in a weight-bearing fashion*. Get your hips over your ankles and your torso over your hips, then walk around a lot.

BACKING UP AND YOUR LOWER BACK

Again, using the images on page 130, check out the movement of the lower back in a forward-hip position. Constant forward motion of the pelvis relative to the backward motion of the torso compresses the lumbar (lower back) region, which changes the angles and loads to the vertebrae, intervertebral discs, and sacrum. What can you do? YOU CAN BACK YOUR HIPS UP. Also, drop your ribs (see pages 31–35) to decompress things from above.

MIND YOUR PELVIS WHILE PREGNANT

The strain-related issues indicated by the arrows in the photo below are often thought of as "common side effects" of being pregnant and are rationalized using physiological arguments (the hormone relaxin is making you SOFT!). I offer a counter argument, a mechanical argument that suggests that how we feel while pregnant is greatly affected by how we carry and use our body before and during pregnancy.

When your hips shift forward, you change where loads are being

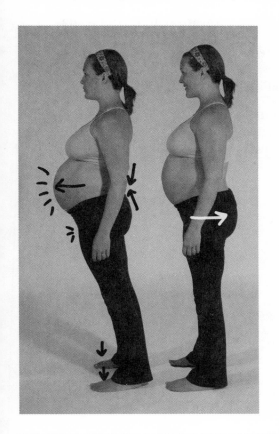

carried throughout the body. A forward pelvis combined with the extra weight of a fetus and its accompaniments can:

- create forces that pull the right and left sides of the pelvis away from each other at the pubic symphysis, straining the pubic ligaments
- press abdominal contents *extra hard* against the inside of the front

abdominal muscles, which can excessively widen the space between these muscles (i.e., create a separation in between these muscles called *diastasis recti*)

- compress your lumbar spine and push down on your sacrum more than necessary
- place more weight on small tissues of the front of the foot than needs to be there

In short, many musculoskeletal issues associated with pregnancy relate more to how weak our bodies are overall, i.e., the added mass from pregnancy easily pulls the pelvis forward with no opposing strength on the back side. The effects of weakness are constantly amplified by the steadily increasing mass of pregnancy, so BACK UP THOSE HIPS (and make sure your footwear allows you to). This will take a load off the parts that are bugging you and place it on the parts that need to work more.

A NOD TO THE SACRUM

The pelvis is made up of three bones: a right and left *os coxae* ("hip bones") and a sacrum wedged in between. The *sacroiliac joints* or *SI joints* are where the top of the ox coxae—called the ilium—and sacrum connect. The pubic symphysis is the joint at the bottom front of the pelvis where the os coxae's right and left pubic portions come together.

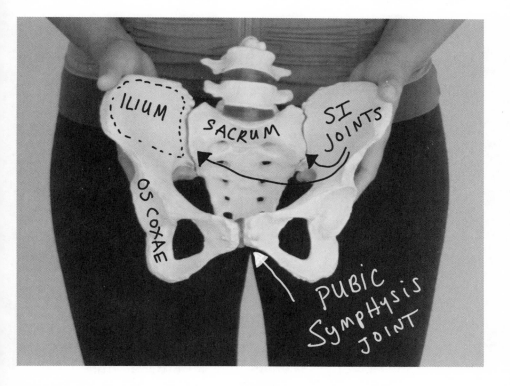

When the pelvis rotates, tilts forward and back, or tilts laterally (as it does during the Pelvic List on page 107), the entire pelvis—that is, all three bones of the pelvis—moves together as a single unit.

But movement also occurs between pelvic parts. For example, the sacrum can tilt forward and backward at the SI joints relative to the other two pelvic bones. This movement is tiny compared to how much most joints can move (although women typically have more SI joint mobility, likely to allow for potential vaginal childbirth). Too little sacral movement (overly stiff SI joints) and too much sacral movement (unstable SI joints) each come with their own physical issues.

The sacrum's hinging movement is called *nutation* (forward tipping) and *counternutation* (backward tipping). Just as with the naming of whole-pelvis movements, sacral movements are named for the direction the top of the sacrum moves.

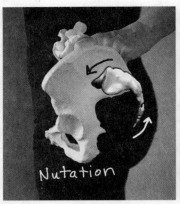
Nutation

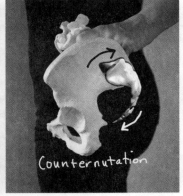
Counternutation

So to recap: The tilt of the whole pelvis is different from the tilt of the sacrum within the pelvis. A pelvis that's tilted anteriorly could have either a nutated or a counternutated sacrum and a posteriorly tilted pelvis can also have a nutated or counternutated sacrum. The position of the pelvis doesn't indicate the position of the sacrum.

Pelvic position affects the way the muscles that attach to the pelvis work and how those muscles go on to impact the joints they control. Hamstrings, quadriceps, abdominals, knees, hips, and the lower spine are all affected—for better or worse—by how mobile the pelvis is and the position it rests in most often. Pelvic position also affects how the glutes and pelvic floor work, and how the sacroiliac joint feels. Part of this relates to the tilt of the entire pelvis and part relates to the tilt of the sacrum within the pelvis. Here's why.

These muscles connect to the *posterior* (back side) of the sacrum:

- the biggest butt muscle (gluteus maximus)
- your upper back muscles (latissimus dorsi, via the thoracolumbar fascia)
- long muscles that run along the length of the spine (erector spinae and multifidus)

Of all these, the glute muscle is the biggest and can be the strongest (but it's often weak and inactive through habit).

Because of where these muscles attach, they tilt the sacrum into a nutated position when they contract (see arrows).

On the opposite, *anterior* side of the sacrum, the muscles that attach are:

- the piriformis (a muscle deep in the hip, stretched by the Number 4 stretch on page 111)
- a *tiny* bit of your iliacus (a hip flexor stretched by exercises on pages 147, 195, and 206)
- one of the pelvic floor muscles (the coccygeus)

When these muscles tighten, they tilt the sacrum into a counternutated position—that draws the tailbone deeper into the pelvic bowl (see arrows).

The nutation/counternutation movement at the SI joints affects the health of the lower back and pelvic floor. We want our SI joints to allow the sacrum to move a bit when necessary but not so much that the sacrum and entire pelvis become unstable. An unstable pelvis can make the simple act of standing and walking a pain in the SI and pubic symphysis joints.

Our pelvic floor is always under a load, so it's always working and creating tension in a direction that counternutates the sacrum (which is the position that offers the least amount of pelvic stabilization). This is why we need glute strength: to maintain a balancing tension on the opposite side. Glutes are part of the structure of the pelvis—the bones and ligaments cannot do it alone.

You also need glutes so the pelvic floor can create leverage. Imagine you want to create a *big* contraction in your biceps muscles. In order to do that you need a weight in your hand, or to hold on to a bar or something that resists the movement. Without that resistance, the contractions can only be small. It's the same with the pelvic floor muscles, only they pull on the sacrum. In order to get bigger forces going on

the pelvic floor, the sacrum needs to be held fast by butt muscles, maybe even pulled gently away from the front of the pelvis (see arrows). If you have an always-working pelvic floor paired with a barely working butt, things aren't going to work as well in the pelvis as they could.

Sacral position is not typically included in descriptions of an anatomically neutral pelvis, but it should be. It's understandable why it's not; sacral position is impossible to see, hard to feel except in extreme cases (childbirth or injury), and not easy or practical to measure internally. Still, the SI joints and sacral lever need a nod. It's important to our spines and pelvic floor parts, to actions like walking, birthing, and bending over and back up again with ease, and to the physical experiences we want to keep on the table.

BUTT WHY THE NEUTRAL PELVIS?

In addition to helping you get up and down, and walk you around from place to place, the largest butt muscles are important to the lower back and pelvic floor. Without strong glutes, the sacrum isn't stable at the sacroiliac joints, and the pelvic floor has a harder time contracting properly to get its job done.

The gluteus maximus contracts when the thigh bone moves behind the pelvis at the hip joint (a motion called *hip extension*). This is why all those 1980s workout videos had people on their hands and knees, lifting one leg up behind them three hundred times in a row. Buns of steel, baby!

Another place hip extension could show up is during walking… although how much it extends depends on the position of your pelvis.

When we walk, one leg moves out in front of the pelvis while the other leg is moving back. The leg moving back is the one carrying our body weight. That backward motion of one leg while carrying our body weight is similar to the 1980s butt exercise, only it's done while standing and probably not while wearing a shiny thong leotard. (Although you do you. #nojudgement)

Butt, in order to get that hip extension/glute contraction combo with each step, your hip joint has to be mobile enough to allow your thigh bone to move back, and your pelvis has to be positioned so that the thigh has the greatest range of loaded hip extension—which means it has to be neutral.

It's easiest for me to explain using the drawing below.

Our limbs can't move just any which way; they are limited in how far they can move in each direction by the shape of the joint—or more specifically, by the shape of the bones that come together at the joint. In the case of the hip, the thigh can move forward until the shape of the hip bone (os coxae) stops it, and the same for going backward. (P.S. This is just the skeleton I'm talking about. Most of us have such stiff muscles we're nowhere close to the range allowed by our bones—our muscles stop us short!)

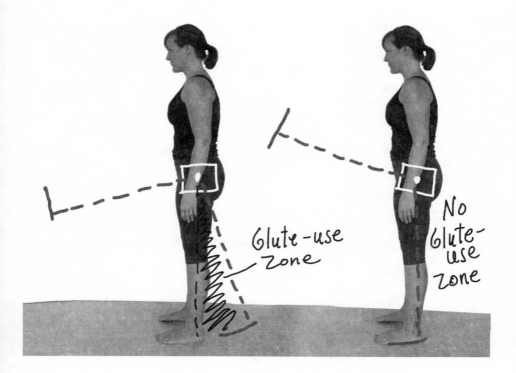

Glute-use Zone

No Glute-use Zone

Let's say your hip has a total range of 100° in the front-to-back direction. How that range of motion lines up with the ground is important. If your pelvis is tucked, that places most of your hip's range of motion in front of your body (see right side of image). You could do a high kick, but then the back of the hip joint would stop the leg from going behind the torso where the butt workout begins. That means the distance over which you can carry your body while walking, i.e., how long your butt can work with each step, or your stride's "glute zone," is short.

So a neutral pelvis (left side of image) sets up the hip joints for maximal butt work. You just have to make sure the hips are loose enough to extend, which is no easy task. Ironically, not only does all that sitting fill up the day so there's no time left for steps and butt contractions, but the way our body adapts to sitting—the cementing of our pelvic tilts—makes the butt unable to work fully when we finally take some steps. This is why we need to stretch the parts pulling on the pelvis (see pages 51, 102, 108, 171, 183, 193), *and* we need to get up more often and walk around.

A neutral pelvis is great for standing and sitting (it sets the stage for better spinal loading and abdominal leverage, and unloads your weight from your sacrum while sitting). But **what the glutes do while you're walking** gives the pelvic floor, which is always contracting under the load of the abdominal and pelvic floor organs, the constant resistance to generate the eccentric force it needs to keep from shortening and helps to stabilize the pelvis and lower back as a whole. Thanks for assking.

HOW FAR CAN YOU EXTEND YOUR HIPS?

Lie face down and get your ASIS and PS on the ground (or, if you've got a belly in the way and/or if you're prego, you can do this standing; just start by aligning your pelvis to neutral).

Without tipping the pelvis (which is tricky because the pelvis will tend to tilt forward), lift your leg as high off the floor (or behind you when standing) as you can—again, without taking the pelvis with you.

Make sure you're not bending your knee (use a mirror), which is common when your hip-flexing muscles are tight.

Measure the angle between the thigh bone and the pelvis to get the number of degrees you can extend (the average hip has the structural potential to go 20° to 30°).

P.S. The amount your leg goes back behind you in this exercise is the amount your leg could potentially go behind you, contracting your butt, with each step when you're walking. I say *could* because in order to use your glutes through this range of motion, your pelvic-listing muscles have to be able to carry you the entire time on the standing leg, and your heel has to stay connected to the ground as long as possible, which—it's all coming together now—requires enough calf muscle length. Which is the inspiration behind that old song, "The Calf Bone's Connected to the Pelvic Floor Bone."

SNAPPING PSOAS

Have you ever gotten up from a sitting position and felt a little *tweaky* in the hip? Heard an audible click when taking your first few steps after being seated, or a "pop" when kicking your leg out to the side?

This is called "snapping hip." Snapping hip is broken down into two categories: *External snapping hip*, most commonly created by the iliotibial (or IT) band, is a snap over the top of the thigh bone, and *internal snapping hip* is where the iliopsoas tendon snaps over a bony protrusion in the pelvis or thigh bone.

Snapping sounds in the hip are created in a way that's similar to the sounds coming from snapping fingers. The volume of finger snap is based on how much you push the thumb and middle finger together. If you barely press them together, there's not much noise. But if you push them together with greater force, you create a bigger snap sound.

A primary characteristic of muscle tissue, or "what makes muscles muscle," is that it can lengthen and shorten. This is what allows our joints to easily change their position, and thus our bodies to move. But when we don't move all that much, when we rarely change positions, our muscles lose their ability to change their position too. Many muscles get shorter and pull tighter, and this is the case with the muscles that shorten as we sit, like the muscles that pull on the iliopsoas tendon and IT band.

The tendon of the psoas and iliacus muscle should glide gently over the pelvis and thigh bone as you move your leg, just like your two snapping fingers barely pressing together and making no noise as they slide across each other. But tight muscles transfer tension to their tendons. In the case of snapping hip, the tendons of tight muscles are pulled into the bone (like pressing your snapping fingers together firmly), the tendons gets hung up on small bony protrusions, and the *noise* you hear (or sometimes it's just a snap you *feel*) is created by the tight rubbing as the tendon passes over. Oh SNAP!

To reduce the force of the snap (and thus the sound), you need to lengthen the muscles. You can pair stretches that move these muscles (pages 51, 86, 102, 193, 195, 206) with sitting less and walking more, which can help get them out of their shortening habits.

When an internal snapping hip moves beyond the noise phase to one that starts damaging tissues due to too much rubbing, there is a surgery for dealing with too much muscle tension called an *iliopsoas tendon release*.

As a geometry nerd, the way this surgery works tickles me. Not the getting it part, but *how* the surgery makes this muscle longer. Now you may wonder—and really I'm just hoping you wonder, because I'm going to tell you anyway—how does cutting something make it longer? Well, while doing laundry years ago, I came up with a way to explain how the surgery works using pictures.

This model begins with an old shirt, not belonging to me. The shirt represents the iliopsoas tendon, and the man with hairy arms is the psoas muscle. The man's arms pull as the psoas muscle does and the shirt represents the iliopsoas tendon. Stretched to the max, it is ready to snap over a bony prominence at any moment.

The surgeon is played by me and my sewing shears. Which my husband didn't know were sewing shears and had been using for the last year as regular scissors, mostly for cutting boxes, which is why I could barely cut the shirt. I mean tendon.

Comparing the pictures side by side, you can see how cutting slits allows the shape to change in a way that lengthens the distance between the top and bottom of the tendon.

The downside is that the structural integrity of a material is determined by the continuity of its fibers. When you take a material under tension and cut a hole, the cut edges can move away from each other, kind of "thinning" the tissue. Cutting into the shirt made it longer, no question, as cutting into the psoas tendon makes that tendon longer—**but it does not lengthen the psoas muscle itself**. And, P.S., I don't think a surgery has this many cuts, I just want you to see how continuity affects shape.

This surgery can stop the rubbing and snapping, but the ability for the psoas to be a strong muscle is compromised because it now has to try to generate leverage while it 1) is still tight and 2) connects to a structure that's shredded in parts.

This approach feels a bit like disconnecting the fire alarm instead of putting out the fire. I see the fact that the muscles aren't working as the fire. But I also see it another way too. If the psoas is tense to the point where every movement of the hip hurts, your *fire* is that pain, not the tension itself.

Luckily there are movements to be done over time that could help, and in a way that preserves strength and tissue integrity going forward.

TRY THIS PSOAS RELEASE EXERCISE

Lie on your back with your legs stretched out straight on the floor, propping your head and shoulders on a bolster or stack of pillows, **with nothing but space underneath your ribcage**. You'll want your ribcage to have space to lower into.

If your ribcage is up in the air when your thighs are down on the ground, then your psoas muscle is not relaxed to a length that allows you to stand fully upright.

When a psoas muscle relaxes, your ribcage should lie on the floor (bottom picture) and not thrust up (top picture) or elevate the knees when it drops (middle picture).

Note: **You shouldn't have to contract any muscles** to get your ribcage down. The idea is to relax into this position without adding more muscle tension. There are no points given for the position itself; it's the relaxed state of the parts inside we are after.

If you find your ribcage is up in the air as in the first picture, just sit (lie, really) with it. There is nothing to "do" to fix it. The only thing to do is close your eyes and scan your body to see if you're holding tension in the torso. You're like an onion, so you'll need to progress in layers, relaxing in waves.

A STRONG, SUPPLE "WHOLE-BODY" PELVIC FLOOR

Every body has a pelvic floor and every body could use a strong one. The standard exercise-type treatment for a weak pelvic floor is Kegel exercises, and a lot of people's understanding of pelvic floor exercises begins and ends with them. But the issue is much more complex (surprise!) from a biomechanical perspective, and the more we know, the more movements we can put into our pelvic-floor toolbox. Although if you're carrying a whole toolbox with your pelvic floor, then maybe spend some time relaxing it.

Some key points often missing from the mainstream Kegel conversation include:

- Not all pelvic floors are weak due to being too stretched out; some will measure as weak due to *too much* tension.
- The butt muscles (and other leg muscles too) are important when it comes to moving the levers of the pelvic floor. An *all-Kegel* or *all-pelvic-floor* approach neglects much of the anatomy affecting a pelvic issue.
- Men also have pelvic floors and can have pelvic floor disorders. All bodies do and can.

- Being pregnant or having had babies is not the only risk factor for pelvic-floor issues; they can also show up in bodies that have never had babies in them.

There are many different issues of the pelvis and each is multifactorial. They are not all *created* by a lack of movement, but they might all be *affected* by a lack of movement. Because we're a sedentary culture, we don't currently perceive and thus investigate sedentarism as a factor, and the exercise treatments we come up with are typically low in volume (minutes a day, for few days a week), which makes it easy to conclude, anecdotally or scientifically, that movement as a treatment doesn't work.

My most important message about the pelvic floor is also my most important message about the entire body. Sedentarism does not work for any part of the body, and pelvic floor problems are strongly influenced by it. Prescribing tiny exercise-type movements like Kegels may help in the short term, but it's the equivalent of prescribing vitamin C in a pill. Imagine if we thought vitamin C only came in pills and never knew that we could get it from an orange!

If we went to a nutritionist and only received supplements, we wouldn't become as well as if we were given more direction on healthier eating, food shopping, and cooking. The same goes for movement. A handful of one single exercise won't take us as far as many movement meals made of various movements.

KEGEL WHILE YOU WALK...

To their credit, a nice thing about Kegel exercises is they can be done anywhere. This (sort of) matches the way the pelvic floor works already.

The pelvic floor sits beneath the weight of all your internal organs and is moved when you move. It contracts in a Kegel-like manner (called a pubococcygeal contraction) to varying degrees when you get up and back down again, and with every step as you walk, with all of the other pelvic floor-supporting muscles working too. However, when you sit atop the pelvis most of the day, and move your body or change your body positions infrequently, the pelvic floor does very little. It has little opportunity to feel and adapt to the load above it. Like any muscle not being used most of the time, pelvic floor muscles lose strength and the joints stiffen as they adapt to not getting much movement. So when you *do* go to move or create a rare, big load on the pelvic floor (jumping, running, laughing, sneezing), the parts can't generate the leverage necessary to deal with this sudden demand.

So, what's the best solution? Should we keep mostly sitting down and just add an exercise (Kegels) that *simulates* how the pelvic floor moves as the body changes position? Or should we work to change the body's position more frequently in a way that *stimulates* the pelvic floor muscles with a variety of loads so they begin to respond reflexively?

I do understand the need for therapeutic/corrective exercise (I teach many of these exercises and I think they are great), and I even see a place for Kegel exercises as a beginning step to moving pelvis-parts more. But a single exercise cannot competently train an area of the body that connects to many other parts and has so much riding on it.

MOVEMENT AS NUTRITION: IS YOUR PELVIS STARVING?

Many people are already exercising or searching for how to move their whole body well. Often, once people start moving their body more, it becomes clear that the pelvis isn't keeping up. If you experience urine leakage, organ prolapse, or pain during exercise, you might wonder, HOW IS MOVEMENT HELPING AGAIN?

In these cases, it's helpful to consider movement as nutrition. Food is good in general, but maybe there are some foods you can't tolerate or maybe you're just taking in too much of one food. Even the most nutritious food will leave you sick if it's all you eat. You need a range of nutrients as they all work together. So, what's your entire movement diet? Is your exercise mode (running, biking, lifting, Pilates, etc.) the only movement-food you eat? Are you balancing your bout of *concentrated movement* with a host of other movement nutrients throughout the day that support and strengthen your parts, including your pelvic floor? Or are you movement-starved from head to toe?

Your pelvic issue might be the result of some small movement nutrient you're missing, a single part that could use some care, but it could also (simultaneously) relate to your pelvis's broader need for more "movement calories" each day.

WHICH PARTS DO YOU NEED TO GET MOVING?

Pelvic ailments require us to think more broadly than the pelvis; there are simply too many moving parts outside of our pelvic anatomy that affect how it works. The pelvis is the junction of many parts, including those seemingly far away. (For example, did you know there are muscles connecting your pelvis to your shins? So movements of your knees and shin bones move your pelvic-floor muscles!) Your pelvis parts are affected by all sorts of whole-body movements—and not only obvious movements like running or jumping, but even more subtle things we don't even think of as moves, like breathing.

This isn't to say that all pelvic issues are a result of a movement deficit, but biomechanical factors often play a role. It's important to understand all the pelvis's moving parts and the parts connected to it when we're troubleshooting what's happening.

There are multiple leverage systems affecting the pelvic bowl; many are listed below.

The bony levers:

- two (one right and one left) os coxae, or "hip bones," each having three named areas: ilium, ischium, and pubis
- a sacrum
- a coccyx
- two femurs

The pelvic floor itself, which is made up of muscles, connective tissue, and nerves:

- tissues of the levator ani (which is made up of the pubococcygeus, iliococcygeus, puborectalis muscles)
- tissues of the coccygeus muscle

There are also muscles not included in the "pelvic floor group" that make up the pelvic walls that seal the bottom of the bony pelvis. These are classified as hip/leg muscles, but the hips also belong to the moving parts of the pelvic floor:

- internal and external obturators
- piriformis and other deep hip rotators (gemellus superior, gemellus inferior, quadratus femoris)
- the "lateral hip muscles" (tensor fascia latae, glute medius & minimus)

These are other large tissues directly connected to the pelvis but also not classified as pelvic muscles, so it's easy to accidentally leave them off the "troubleshooting my pelvis" radar:

- the gluteus maximus
- latissimus dorsi (these are one of the reasons TIGHT SHOUL-DERS can affect the pelvis!)
- thoracolumbar fascia

The gluteal muscles pull the sacrum in opposition to the pull created by the pelvic floor, as diagrammed on page 139.

But consider what the pelvis does: it bears the burden of the abdominal and pelvic contents, and it helps facilitate breathing and coughing (both helpful when it comes to keeping the lungs clear). And consider how much position influences weight, pressure, and leverage. It becomes obvious that we must expand our list of levers and pulleys affecting the pelvis; we have to include those that tilt or tuck the pelvis. So, here are even more muscles (and their nerves and surrounding connective tissue) that affect pelvic position or loads:

- rectus femoris (one of the quadriceps that connects the pelvis to the shins)
- hamstrings
- adductors (muscles of the inner thighs)
- psoas major (via flexing the lumbar spine)
- iliacus
- quadratus lumborum
- muscles of the abdomen (abdominals, diaphragm)

When we acknowledge how many other parts are involved in moving the pelvis and its parts, then the list of exercises we need to do to maintain our pelvic health can quickly get longer and longer.

HOW TO START MOVING YOUR PELVIS PARTS

Clearly we need to move our entire body more, and there are many parts in your body to get moving—so where to start? The short answer is, I don't know the right answer for you, mostly because I don't know who you are, how you're moving now, what your specific issues are, and what movement protocol will fit into your life. That said, I've worked with many people on their pelvic floor issues, and most of them have benefitted by working in this order:

1. Small corrective exercises
2. Everyday alignment adjustments
3. STACKING your movement
4. Walking more
5. Hanging

Note: You don't have to complete a category before moving on to the next (you can start walking more now, too!); I just want to stress that getting your individual parts moving is the first step to attaining the larger movements to benefit your pelvic floor more directly.

1. Small corrective exercises

Whether you exercise a lot or not at all, it's possible to have pelvic-affecting parts that don't move. When we go to move parts that have been stiff for a long time, they resist this change so we tend to move parts around them instead. This avoidance happens without us realizing and can be hard to detect, which is why I teach movement with such specific form. You might already have a favorite calf stretch, or have been stretching your hamstrings for years and think all your parts are moving well, but without using exact form to measure, it's easy to inadvertently leave "sticky parts" unmoved. It's possible you don't need new exercises, just more discerning and better-trained eyes.

For example, I've seen multiple clients (scores of them, even) who have been told their hamstrings aren't tight or affecting their pelvic floor issues because they can easily bend over and touch the floor. But you can bend over and touch the floor by using excessive lower back movement and no hamstring movement at all. Being able to touch the floor doesn't mean your hamstrings are mobile; to discern your hamstring mobility, you have to watch how far the pelvis can move over your thigh-bones. If you're watching your hands instead of your pelvis, it's easy to miss the fact that your pelvis is bound to your thighs.

The lessons in this book will help you see more clearly how individual parts can move. Getting your calves, hamstrings, hip flexors, waist, and diaphragm moving better is where to begin. See: ALL THE EXERCISES IN THIS BOOK.

2. Everyday alignment adjustments

Once you've got your pelvis-affecting parts moving a bit (which is sort of all of them), practice the simple adjustments to body position that can quickly change pelvic loads throughout the day. Back up your hips, find a neutral pelvis, make sure your ribcage isn't lifted, and make sure your shoes aren't pushing your pelvis around. All the exercises in this book (even the ones for the feet and arms) make it easier to adjust your position, and adjusting your position actually uses your parts more all day, making the exercises easier and, over time, less necessary. It's cyclical, and you might find that exercises and adjustments you can't do right now become more available over time by doing the ones you can.

3. STACKING your movement

Our bodies need a lot more movement and more diversity of movement than they're probably getting, but we've set up our lives in a way that allows no more time for movement! Good news: You can start layering movement into other things you're doing (I call this "stacking movement") by making a few simple changes.

Sitting on the floor can give you the time to do whatever else you're doing (working, watching TV, reading, hanging out) while also doing leg-moving exercises. I'm not going to say you *can't* take a Zoom conference in Psoas Release (page 147), but maybe opt for a V-sit to move your hamstrings while on camera, or if you're in an actual office, just cross one leg over the other while in your desk chair in a Number 4 position (page 111) and move your piriformis *on company time*. Stack your evening reading with Legs on the Wall (page 172).

4. *Walking more*

Walking is basically the best thing you can do for your entire body as it's free, functional, and it moves a lot of your parts at once— pelvic floor included. Even before you've started all the little exercises above, walking is an excellent low-impact way to move. Once you've checked in with your parts and got them moving more, you can refine your walking so it's even more nutritious. First, tune in to how much you are walking. Break up bouts of sitting on your pelvis with walking your pelvis around, and using your pelvic floor and its friendly lever-mate, the butt, at the same time. Aim to increase your walking by ten minutes a day, then slowly add more over time.

Take a longer walk each day as you can (stack it with running errands or visiting with a friend on foot), but also look for places where a short, 3–5-minute walking session can fit in. If you think of walking as your pelvic floor activity time, then every minute you're breaking up a sedentary position counts.

Check out your walking-movement diet. Are you always on flat-and-level sidewalks? Look for bumpy terrain, hills, and steps that move your legs (and pelvis) more. Carrying things while you walk will expand your abdominal-movement diversity.

5. Hanging

In short, the movements and tensions of your shoulder girdle (and abdomen) are important when it comes to intra-abdominal pressure being high, which is part of what can be pushing down on the organs of the pelvis. Also, the tension in the latissimus dorsi—the broad muscles of the upper back—contributes to the stabilizing movements of sacrum (which helps the pelvic floor by providing an opposing force). In shorter short, weak arms end up affecting the pelvis. Work on mobilizing your shoulders and strengthening your arms in general so you can take that next step and hang from a branch or bar to stretch and strengthen them further (see how to create a hanging station on page 66).

MOVEMENT PERMACULTURE FOR THE PELVIS

Once you've taken a survey of the "sticky spots" in your body and figured out how to isolate them, you can add your stacks, **restoring the pelvis while you restore movement back into your life**. Consider the movements that can make up a lifetime: getting up and down, walking, running, bending over, carrying stuff, climbing, and even sitting low to the ground or squatting to rest. They *all* hold the potential to move your pelvic parts more (and they can also facilitate other, non-exercise benefits as well). **This is movement permaculture.** By tending to the soil of your life, you can simultaneously reap the "healthy plant" of a pelvis that works better for you.

MENSTRUATION IS A MOVEMENT

Menstruation is a movement. It's a movement that until recently, humans have done while moving their bodies around a lot. When we think about the inner workings of our physiology, we should bear in mind that one of the most novel things about our environment is that it is relatively movement-free.

There's now a tremendous number of menstruators (hard to pinpoint when considering all populations, but it appears to be at least 50 percent) who have *primary dysmenorrhea*. Dysmenorrhea is a pain, literally. It's a painful period, ranging from light to severe, that can also include vomiting, diarrhea, headache, and fainting. The "primary" means that the dysmenorrhea is not occurring simultaneously with another known pelvic pathology. (*Secondary dysmenorrhea* can be created or affected by issues like endometriosis and fibroids that can cause extremely painful periods, and I know that people suffering from those conditions have likely heard a million "just do this" solutions, and I'm not trying to offer that here, although I hope that movement might relieve some part of your pain.)

I used to have painful periods. Through my teens and twenties, during the first two days of my period, always starting at night, I'd wake up with achy legs and feet. Then I'd be curled up in a ball, feeling like I might throw up. To cope, I'd take about four Tylenol to get me through these two days, then I'd be back to normal.

Our collective period pain requires a tremendous amount of medicine so we can endure this foundational natural human movement. I'm not saying that to shame people needing medicine—I've been there—but to highlight the severity of the issue: that we humans are having major problems with a *baseline* movement.

MOVING THE SEDENTARY PELVIS

When we think of our pain, the sedentariness of our bodies is not on our radar; a mostly sitting pelvis and uterus are the norm, and so are medicalized periods. I was a healthy, exercising fool back when my periods used to wipe me out, which is why the difference between exercise and movement is key here—my whole-person fitness exercises weren't the movements that ultimately got me out of pain. (This is because even exercisers in our culture are sedentary in terms of how much time they spend unmoving!)

When I was in graduate school studying movement, I focused on the mechanics of pelvic disorders. It was there that I put painful periods in the same category as other pelvic issues, like incontinence or sacro-iliac joint instability and pain—and looked at how all these could be symptoms of sedentarism manifesting in the pelvis (which isn't to say the movement environment is the only factor, but that it certainly needs to be addressed).

There were two things I'd always noticed about my dysmenorrhea. The first was that my uterus felt cold. In fact, I used to sit over the toilet and pour warm water between my legs for relief if a bath wasn't available. The second was those aching legs and feet the night before my period.

I dug into research investigating how the uterus worked, and it was there I found that people with dysmenorrhea experience decreased blood flow to their uterus (and those with severe symptoms have decreased blood flow in the main artery as well as the smaller branches). That made me think about local musculoskeletal movements and how those influence vasodilation and blood flow to those areas.

Then I went to an artery map to see where the uterine arteries begin and found that they branch off the internal iliac arteries, which branch from the common iliac arteries. Then I thought about which local movements of my skeleton would get my skeletal muscles contracting and opening (vasodilating) those arteries. The answer was movements of the lumbar spine and of the hip joints.

At that point I was always moving my legs a ton, being a runner, a (terrible) triathlete, and a, wait for it, STEP AEROBICS INSTRUCTOR AND YES IT WAS AN AMAZING TIME IN MY LIFE.

But despite hours of movement a day, my hips were very tight. Although I moved my legs a lot, I wasn't actually moving *all* of my leg parts—my hip joints didn't move much at all (I didn't have much hip extension and my groin was tight). So I started working on a set of exercises specific to improving the movements in the hips and pelvis, and I really concentrated on form. There's a ton of great exercises out there, but so many of them are done inexactly. While they get a whole person moving, they're hardly effective in increasing the movement in certain spots.

I wound up with a set of exercises that helped me move sedentary spots around my pelvis. It really helped! They took me out of severe pain to mild discomfort.

- Calf Stretch (page 217)
- Double Calf Stretch (page 171)
- Lunges (page 205) for my hip flexors
- Forward bends (page 184) and Strap Stretch (page 182) for my hamstrings
- Spinal Twist for the muscles in my lower waist (page 102)
- Legs on the Wall for my inner thighs (page 172)

General stretches for periods aren't new. You can find them in 1950s popular health titles and there's clear scientific research on how they can reduce period pain and the need for pain medication. I've always found that thoughtful movement supports my body during times of menstruation (and now pre-menopause, for what that's worth), but general movement wasn't enough—I've found I need targeted stretches and movements for my core and hips to care for myself most.

WHERE DO THE FEET COME IN?

A few years later I started wearing minimal shoes and took myself on a twelve-mile hike wearing Vibrams—shoes that are pretty much like wearing socks. I won't lie, this hike was really hard because my feet had never moved that long over really hard and lumpy ground, and I was limping at the end. But, BUT, the period that came a few days later arrived silently. No aches, no pains, *not even mild ones*. I couldn't believe it.

Ten years later, after wearing only minimal footwear, I rarely have any discomfort, although I still get a tiny bit leg-achy just beforehand when I haven't been moving as much (I never remember why—I'll just be wandering around my house, asking for foot massages, and my husband points out that it must be period time). For my forty-fourth birthday I challenged myself to walk ten miles a day for forty-four days, and the two periods I had during this time were like the time I walked twelve miles in Vibrams—ninja-stealthy in their arrival. Maybe it wasn't the minimal shoes at all, but walking a ton in them. Of course, a high volume of walking with good hip movement is going to bring more blood through the iliac arteries. Although I don't know the exact mechanics of how or why foot, ankle, and leg movements affect the way my period feels, I know they work, so I do them.

WE HAVE TO MOVE. PERIOD.

We can't ignore our sedentary culture when it comes to understanding and detailing how our body works. Maybe your painful periods are just sedentary ones, like mine were. To test it, you can try moving more of your pelvic parts as well as your whole person more.

And that is all I have to say about periods.

LOW-HANGING FRUIT

There are a lot of movements going on in the pelvis, including the lifting and lowering of the testicles.

The *cremaster* (from the Greek verb "to suspend or hang") is the muscle responsible for keeping the sperm-generating process underway by regulating temperature in *le sac*. The primary theory as to why these precious jewels sit so vulnerably outside the body is that it's because the sperm-generating process is very sensitive to heat. The cremaster is constantly lowering the testicles when they're too hot and pulling them close to the body when it's too cold for the sperm cells. The cremaster is like Goldilocks, looking for a temperature that's *just right*.

The cremaster also elevates the testicles in dangerous situations (like tennis), elevates them during erection (which might be part of what makes an erection), and ejaculation.

Comedian Mitch Hedberg said this: "My belt holds my pants up, but my belt loops hold my belt up. I don't really know what's happening down there. Who is the real hero?"

I think of this joke when I think of the cremaster muscle because the cremaster originates from two of the abdominal muscles: the internal oblique and the transverse abdominis. So what's actually holding up the balls? Is it the cremaster or is it these abdominals or is it both, with the cremaster only able to do what it can based on what the abs are doing?

When it comes to functions of the penis and testicles—parts that aren't included in the musculoskeletal system—how do their functions (erections, ejaculation, making sperm) depend on the positions of all the other body parts? How is their function impacted by the positions we spend most of our time in?

The movement of abdominal muscles is affected by pelvic position, sacral position, and the resting tension within the abdomen (see Abdominal Release on page 41). Pelvic and sacral position are affected by muscles on the front and the backs of the thighs, your glutes, and your pelvic floor. Are these muscles tight? Loose?

My point is you've got a lot (BALL MOVEMENT AND ITS FUNCTIONS) riding on many non-testicle muscles getting the movement they need. In case you needed another reason to prioritize moving your body well.

And that's all I have to say about testicles.

HEAD-TO-TOE PELVIC FLOOR MOVES

Just as your body dwells in the ecosystem of the planet, your pelvis lives within the ecosystem of your body. Thus the state of your pelvic floor depends on the state of all the parts that attach to it. Remember when I said pelvic-floor disorders are sometimes caused by pelvic floors that are already *too* tight? In that case, Kegel contractions are the opposite of what's needed and are ineffective.

There are ways of creating a healthy, responsive pelvic floor that can tense *and* relax as necessary (i.e., depending on what you're doing)— by using more of the body parts surrounding and connecting to the soft tissues of the pelvis.

If you enjoy doing Kegels, that's fine, but you can also expand your "healthy pelvis" routine to include a few other exercises that hit different areas of the pelvis as well. And bonus: these moves don't only change the loads to your pelvis, they also change how you're loading your feet, knees, hips, and low back...all day long.

HIPS OVER HEELS

Adjusting your hips back is a simple way to instantly make over the loads placed on the pelvis and pelvic floor (see image on page 129). Fit this move into all those times you find yourself standing: in line, at the kitchen counter, in front of a computer, on the sidelines of a soccer game, etc.

DOUBLE CALF STRETCH

Pelvic issues are prevalent in our culture and so is a ton of sitting. One way to strengthen your pelvis is to simply get up out of your chair more often. Once you're up, you can try this next exercise using your

chair—it targets muscle tension you might have in your calves and hamstrings, which can pull on your pelvis.

- Stand facing the seat of a chair with your feet pointing forward and legs straight (no bent knees!). Bend forward until your palms rest on the chair (see image previous page). If you can't reach the chair without really bending the knees or rounding the back, add a pillow or stack of books to the seat until you can, or move to a counter or desktop.

- Once your arms are supported, relax your spine toward the floor, and lift your tailbone toward the ceiling. Don't force your ribs to the floor or arch your back—just relax the spine to the ground as much as you can.

- The more you lift your tailbone, the more you'll feel the tension down the back of the legs. Hold this stretch as long as you can, up to a minute, and remind yourself to take more breaks from sitting (a perfect time to do this exercise!).

LEGS ON THE WALL

Tight groin muscles can limit the motion and strength of the pelvic muscles (or perhaps it's the other way around). Try this next exercise to see if you can introduce new movement into this area of your body.

- Sit sideways to the wall and rotate your body to get your legs straight up the wall, lying on your back.

- Back yourself away from the wall until your pelvis can relax into a level position. If it's tucking (i.e., your waistband is pressing into the floor), scoot back until there's a little space under your waistband.

- Keeping your legs straight, relax them away from each other until you feel a stretch in the inner thigh.

- This move can be a doozy, so come out of it as necessary and back into it when you feel ready.

RECLINED SOLE-TO-SOLE SIT

- Lie on your back on a pillow, with your knees bent, feet flat on the floor.

- Slowly drop the knees away from each other, keeping the soles of your feet touching.

- Place pillows under each knee to support you in this position as necessary (note that you might need more pillows under one leg than under the other). Lower the support bolsters as the position becomes more comfortable.

Relax! I'll often do my morning or evening reading in this position— making my reading time great for my mind *and* my hips and pelvis. Once you've done this one for a while, you can vary the move by sliding your feet farther from or closer to you.

BUTT BUILDER

Many are surprised to learn their butt strength matters when it comes to pelvic health, but I've repeatedly found that pelvic floor function is enhanced by strong glutes—which makes sense as they're connected to each other via the sacrum. There are many ways to strengthen your glutes, but this is a fun exercise you can work a few times into each day.

- Stand barefoot in front of a wall and place your hands on it.
- With your hands still touching the wall, walk back and lift your left leg so all your weight is on your right leg.
- Keeping the right leg straight (make sure you're not bending your knee!), shift your weight back into your heel. The wall is there to help—use your hands on the wall to help you find the position.
- Lower the left (floating) side of the pelvis toward the floor—this will engage your glutes, especially if your weight is back and your knee is straight. Again, use the wall to help you find this position!

Over time, the muscles of your pelvis, hips, and legs will become stronger and you'll be able to use your arms less. To make this move more challenging, reduce how much you're using the wall a little bit each time.

CHAPTER SIX

YOUR HIPS, LEGS, AND KNEES

" The floor can do wonders for your hips, shoulders, and spine. It can help you build bone and decrease your blood pressure. But you have to *get down to it*. You need to *get down on it*. And then, of course, you need to get back up."

A USER'S GUIDE TO THE HAMSTRINGS

L et's start with some quick hamstring anatomy tidbits:

1. The hamstrings are a group of three separate muscles: biceps femoris, semimembranosus, and the semitendinosus.

2. The hamstring muscle group is *biarticular*, meaning these muscles pass over two joints. They pass over both the hip and the knee joint, which means they can affect both of these joints. Hamstrings essentially connect the pelvis to the shins (see left image).

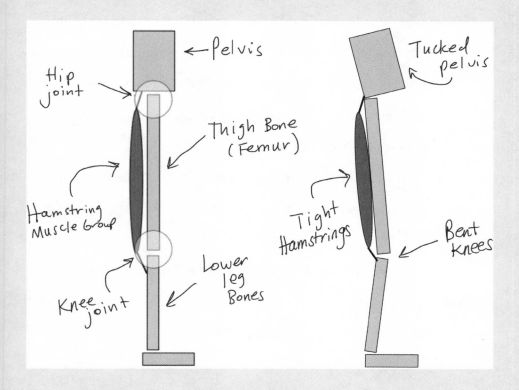

3. When flexible enough, the hamstring group allows the pelvis, thigh, and lower leg to be positioned in anatomical neutral. When not, they can tuck the pelvis and bend the knee while standing (see right image on previous page).

4. The hamstrings are commonly listed in textbooks as knee flexors (muscles that bend the knee), but this muscle group is used more often as a hip extensor (muscles that pull the thigh backward relative to the pelvis) during walking.

5. The backward motion of hip extension is the most efficient way of propelling the body forward when walking (think of an ice skater pushing back against the ice to move forward), but that doesn't always happen. Because of the way bodies adapt to modern environments and to very little daily movement, gait patterns have sort of reversed. Instead of using our glutes and hamstrings to smoothly carry us forward, many of us walk by lifting a leg in front of us and *falling* forward with each step.

6. When you sit in a chair or tuck your pelvis under, the hamstring muscles' attachment sites move toward each other, shortening the muscle. When you sit and tuck often, the hamstrings get used to being shorter and start resisting attempts to lengthen fully.

7. When the hamstrings are adapted to this shorter state, they can hold the body in a way that prevents the knees from completely straightening or the pelvis from untucking when you're standing.

(I write a lot about the pelvis and pelvic floor issues, and about how often the hamstrings are overlooked when it comes to their impact on the mechanics of the pelvis. Hamstring and other hip movements need to be part of a pelvic floor exercise program.)

WHAT MAKES HAMSTRINGS SO TIGHT?

There are many, many unique situations that can affect human tissue, but consider some of the daily position habits we likely have in common:

1. We sit hours and hours a day, all the years of our life, meaning we practice a disproportionately large number of hamstring-shortening movements compared to any other leg motions.
2. Many people have habitually tucked their pelvis for decades because they were taught to do so (for good posture, better manners, etc.).
3. Most conventional shoes have a heel, which positions the foot slightly downhill. The body compensates for the change in geometry underfoot by slightly bending the knees or tucking the pelvis—and both movements shorten the hamstrings.
4. We don't move very much at all, so these muscles never do much work, and when they do, that work doesn't use the full range of motion of these parts.

AND NOW...HOW TO STRETCH YOUR HAMSTRINGS

If you want to change the state of your hamstrings, including their length, you have to move them more. You also need to make sure your hamstring movements, including your stretches, use good form, otherwise the movements won't achieve what you intend them to. Specifically, you want to make sure that you're moving the attachments *away from each other*—not moving both of them in the same direction.

When trying to stretch your hamstrings, make sure you're not bending your knees. Bending your knees brings the hamstrings' attachments back toward each other.

Similarly, don't tuck the pelvis while stretching your hamstrings. Tucking your pelvis moves the attachments toward each other.

To recap: When you stretch your hamstrings, you have to pay attention to the position of your knees and pelvis. Keep them moving away from each other so you don't spoil the stretch. Here's a simple move to practice watching your pelvis and knee position:

Strap Stretch

Lie on your back (or bed, if you can't get to the floor) and wrap a strap or belt around the foot. Then, straighten the knee and untuck the pelvis. You might find you can't do either or both. In this case, lower the leg until you can.

Toes toward the floor

Leg straight →

Arms work, legs relax

Leg Down

Because the hamstring passes over the knee and hip joint, tucking the pelvis or bending the knee once you've started the move takes the stretch out of the exercise.

You'll know the pelvis has tucked if the bottom leg lifts away from the ground. To monitor whether that happens, place a hand towel under the back of the thigh on the ground, and keep it pinned down as you're stretching your top leg. You may need to lower the stretching leg almost all the way to the floor to anchor the other thigh and therefore move your hamstrings well.

It's often tempting to think the goal is to move our limbs as high or as far as we can, but in most movements the aim is to move the smaller parts in between. In a hamstring strap stretch, a really high leg with a bent knee and a tucked pelvis provides very little hamstring movement.

HOW DO YOU BEND FORWARD?

Forward bends are a great way to move your hamstrings, but you have to know which parts to watch, because it's possible for your hamstrings to sneak out of this stretch when you bend forward.

How do you reach forward to touch your toes without moving the hamstrings? Not everyone can do this, but if you've got a really mobile lower spine, it can bend so far forward that your hands touch the ground without your pelvis—and thus hamstrings—moving (middle image below).

To move your hamstrings in a forward bend, you need to tip the pelvis forward (right image), which moves the pelvic hamstring attachment away from the hamstring's attachment below the knee. You can think of your pelvis as a bowl full of soup: tip it forward to pour out the liquid in front of you.

Bending your knees while tipping the pelvis forward takes the stretch away from the hamstrings, so focus on keeping your knees straight to see clearly how far your hamstrings can move.

Now it's possible, even probable, that your hamstrings aren't flexible enough to let your pelvis go very far. If you want to make sure the backs of your legs get the stretch they need, let the movement be small. Stop the forward bend motion once your pelvis can't tip forward any farther, and hold it there for a stretch before standing back up.

Working on a small, pelvis-only forward bend that concentrates the movement on your hamstrings is a good exercise by itself, but you can also add the hamstring movement when you bend over to touch your toes or during a sun salutation in an exercise session. Just go as far forward with the pelvis as you can—hold for a second to make sure your hamstrings are participating—and then continue the rest of the way forward by bending the spine but continuing to keep the pelvis fully tipped.

When your hamstrings and spine are tight, your knees might also want to bend. If they do, just be aware that this movement isn't arbitrary—the need for it relates to what's happening in your hamstrings. If knee bends are showing up in a lot of your "straight leg" stretches or exercises, make sure to do at least a few repetitions without letting the knees bend (remember, you'll likely have a smaller range of motion) to get the hamstrings some of the movement they need.

The key here is to make sure you know the difference between moving forward at the hips and moving forward at the spine. You need both movements, but to keep them both, you have to practice both, which means you need to pay attention.

FIX YOUR DOWNHILL KNEES

Our family recently hiked with another family from up high in the Olympic National Park to the town below—4,230 vertical feet over 8 miles (or 9.5 if you want to take the GPS's word for it). We always do a couple of challenging family "thru-hikes" each summer: point-to-point hikes that add some other element of sticking with the challenging walk over time. A one-day kid version, if you will.

This was no easy trail (and in many spots, it wasn't a trail at all); most steps required mindful foot placement because it was so rugged. But our two families made it. This was the first time I'd ever done so many steep downhill miles without any flat or uphill terrain to break it up. Two days later, muscle soreness was pinging me in three areas on each leg: glute medius, lateral quad, and soleus.

WHY THESE MUSCLES?

Going uphill is hard. Leg muscles have to do a lot of work to lift up the weight of the body, which requires the heart and lungs to move more. Lowering the body doesn't take as many heart and lung motions, because you're being pulled down the mountain, which is easier in one way but harder in another. Declines—especially steep ones—require the body to "put on the brakes" to keep itself from tumbling down.

The areas I felt after my downhill trek are where the body's "downhill brakes" are located.

Let me draw it out for you in a super clear way that couldn't possibly be misinterpreted.

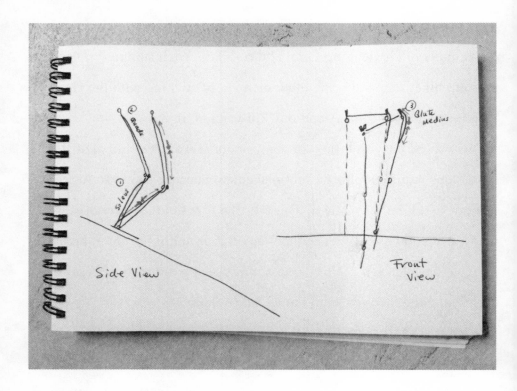

1. The SOLEUS muscle is the deepest calf muscle, and it connects the heel to the shin. When taking a controlled downhill step, the soleus lengthens (**thin arrows**) despite the fact that it's simultaneously contracting (trying to get shorter; **thicker arrows**) so you don't drop down the step too fast. When muscles work in the opposite direction to their net movement, it's called an *eccentric contraction*.

2. Similarly, the QUADS on the front of the thigh must get longer (**thin** arrows) to lower the body, but they must also work opposite to that motion to act as brakes (**thicker arrows**).

3. The front view makes it easier to see that the GLUTE MEDIUS (one of the lateral hip muscles) has the same challenge: to get longer to lower the other side of the pelvis, while also trying to tighten in order to control that motion *eccentrically*.

There are ways to get downhill that don't use the parts noted above. When ankles, knees, hips, and waist are stiff, gait patterns get creative. People with stiff parts can still log many miles a day or even multiple hikes a week; however, with time, these creative alternative patterns take their toll.

WHY CAN KNEES HURT WHEN YOU'RE WALKING DOWNHILL (OR DOWN THE STAIRS, FOR THAT MATTER)?

1. Your knees are doing more than their share of the work

The first thing I have folks check on their downhill gait is "are you using your knees *and* hips, or just your knees?"

One way to save the knee from having to do all the bending work to lower the foot to the ground below is to make sure you're also lowering the hip when you're going downhill. This involves lots of hip muscle use, which helps keep your mass from bombing down the hill.

If you **list and shift the hip**, you create a "slalom" for your center of mass—moving it side to side as it goes down—slowing you naturally (below right). Using your lateral hips to lower and slow your body down with each step more means you don't have to ride the quad brakes so often or so hard. Your knees *can* do all of the work (and probably have been for some time), but they don't *have* to. You have other downhill parts you might just not know how to use, and by the way, your knees called and would like you to learn about them ASAP.

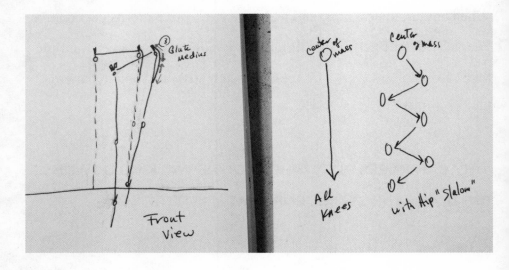

The Pelvic List exercise (see page 107) is key to downhill gait, because this is how you can train the gluteus medius to work specifically while walking (i.e., on a single, standing leg). You can also practice stepping down off a block to prepare your soleus and glute medius to take the workload off your quads.

Remember to do the slalom list-and-shift of the hip while you hike. Even if you haven't practiced it a lot, you can use it to change your gait on the trail and give your knees a rest.

2. Your kneecap (patella) isn't tracking in the "patellar groove"

Going downhill works your quads and thus tenses them. The patella is embedded in the quadriceps' tendon, and if your quads are tense and you're walking downhill, they press the patella deeper into the knee joint than they would on uphill or flatter terrain. OUCH!

Our knees have space for the kneecap, but the kneecap can get pulled out of that space (usually it's pulled sideways) as a result of your all-time movement habits and patterns. Going downhill when the kneecap is out of its groove increases the pressure in the knee. P.S. This can also be why kneeling upon or deeply bending a knee hurts.

My own knees have a tendency to get achy when not cared for, so when I hike I always bring these exercises along to do before, after, and during hiking or backpacking trips: Calf Stretch (page 217), Pelvic List (page 107), Strap Stretch (page 182), and Spinal Twist (page 102). These correctives definitely travel well, and then, so do my knees.

I hope this takes some of the mystery out of why downhill/down-steps walking often makes for achy knees. The good news: it's most often a situation you can do something about.

QUAD STRETCH MAKEOVER

How many quadricep muscles are there on each leg? The answer's in the name: there are four (*quad*riceps!). Three of them run between the thigh bone and the lower leg, but one of them—the rectus femoris— runs between the pelvis and the shin.

Minding your pelvic position when stretching your quads is super important, because if you don't, you're likely to be missing one of them (and it's a big one). If your lower back hurts when you go to stretch your quads, it could be due to a too-short rectus femoris, which can simultaneously tip the pelvis forward when you go to bend the knee for the exercise. As you have already learned, tipping the pelvis forward is one way to compress the lower spine. (Pop Quiz: What's the other way to compress the lower spine that doesn't involved the pelvis? Rib thrusting! So make sure to check your rib cage position in a quad stretch too.)

Here's what to do:

Stabilize your pelvis before you grab the foot or ankle behind you. Tucking your pelvis will keep it in place. Of course, your quads might be so tight you can't reach your foot. In this case, place a strap, belt, or scarf around your foot to help you reach it.

Once you're in the stretch, make sure your knees are together (the stretching knee will sneakily try to escape by moving away from the

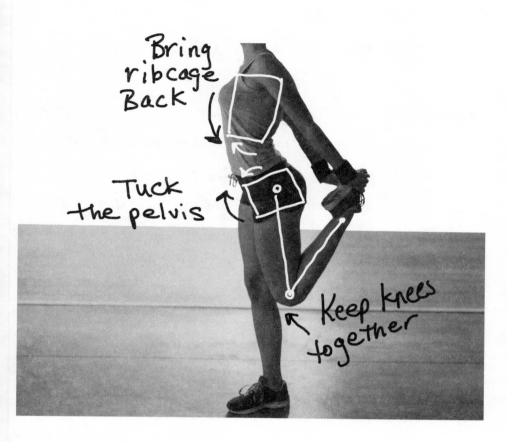

other knee, either out to the side or in front, because that reduces the
stretch too). Then re-tuck your pelvis to make sure you're getting the
stretch where you want it.

I'm not sure if you'll like me or hate me after learning this way of
making the stretch harder. Either way, paying attention to your position
makes this exercise do more of what you wanted, so...you're welcome.

KNEE LOVE AND A LOW-LOAD LUNGE

To reduce knee pain caused by tight quadriceps pulling the kneecap (patella bone) deeper into the knee joint, stiff bits need to be softened up. You might even have been told that it's important to stretch your quads, and that's true. A simple lunge is a great way to start lengthening the iliacus, the psoas, and the four muscles that make up your quadriceps, and **if you mind your pelvis and keep it from tipping forward**, a lunge can really move the quad muscle that's often problematic: the rectus femoris.

But how can you do this great stretch if getting onto your knees or even deeply bending them further presses the kneecaps deep into the joint, causing MORE OUCH?

TRY THE NO-LOAD LUNGE

Grab a chair—an armless chair, a low stool, or something like a piano bench works best.

Sit on the seat sideways so your hips (and most of your body weight) are supported by the chair but your legs are able to move to make the lunge shape in front and behind you. Only half of your butt will be able to sit on the chair.

Once you're in the lunge shape—look, Ma, no pressure on the knee!—all the other lunge position rules apply:

To stretch the muscles down the front of the thigh, tuck your pelvis under (tipping the top of it back towards your spine). To increase the stretch, walk the back leg behind you a little bit more if you can, all the while focusing on not letting the pelvis tilt forward. Moving your leg back while keeping the pelvis still will stretch down the front of the thigh, lengthening those muscles you want to target. Repeat on the other side.

THE BEST PIECE OF EXERCISE EQUIPMENT YOU'RE NOT USING

I see I got your attention.

Now you're wondering what I could be talking about. The treadmill in your guest bedroom with the clothes on it? The roller blades in your closet? The semi-inflated physio ball in the backyard, or the weights in the garage?

It's none of those! This one-size-fits-all, miraculous exercise tool is…your floor.

Really!

The distance between your standing pelvis and the ground seems like one of the hardest to travel, but it's always worth the trip. The floor can do wonders for your hips, shoulders, and spine. It can help you build bone and decrease your blood pressure. But you have to *get down to it*. You need to *get down on it*. And then, of course, you need to get back up.

Our body is made up of many levers and hinges. A big function of those hinges is to get us to and from the ground. Of course, when we spend a lifetime *not* using our hinges to get there, those hinges stop moving well (see: *move it or lose it*).

We quickly find ourselves trapped. When you can't get down to and up from the floor, then you're not going to try to get down to and up from the floor. The experience of something being physically challenging seems like a flag for "you need to not be doing this." It makes sense to keep building up the height of our chairs, beds, and toilets. Our hinges don't bend, so how else can we function?

Movement is one of those weird phenomena for which the solution to not being able to do it—a particular move, for example—is to just keep working on it. Stop when you need to, bolster when you need to, reduce the load as necessary, improve your form, but also, *just keep trying to do it.*

There are many reasons people don't get onto the floor, but almost all of us start out being able to do so. Probably the most practical reason we stop is that we're born into a culture that builds all their seats halfway between standing and the ground and then places these seats everywhere. Your butt has been blocked by the hundreds of cushioned seats and porcelain thrones in your castle.

To start moving your body's hinges more often, just change the way you relate to the floor.

Use the floor for sitting. Our hips and knees don't stand a chance against all those chairs. If you want to start using the floor more often, put a few cushions on the ground where you want to sit and ditch the chair. You don't have to go all the way to the ground at first. Just sitting lower than your regular sitting height will start to loosen your hinges.

Get down onto the floor and up again every day, just to make sure you don't lose the ability to. If you're a movement teacher, have your clients practice the functional task of climbing down to the ground and then back up again. Take turns getting into a face-down position or a face-up position before standing back up. Have a chair nearby, because when the legs aren't strong enough to carry your body weight up and away from the floor, the arms can do some of the work by pushing down on a chair seat.

Want to get your heart rate up? No need for fancy cardio equipment! Begin by standing and then get down to the ground, ending up face down. Climb back up to a standing position and climb back down, ending with your face up. Repeat this cycle five to ten times. It's more fatiguing than you think. Bodies are big weights. Getting up and down from the floor is like a "whole-body bicep curl." As you get down, you lower the weight, and you lift it back up when you stand up. But getting up and down does more than move your heart and lungs; it also moves a lot of your body hinges through a big range of motion—and not all cardio exercises do that!

There are some common excuses for not getting onto the floor that have nothing to do with strength:

My floor is too dirty.

Clean it. When it comes time for your annual doctor visit, how are you going to feel knowing you *could have taken better care of your spine* but didn't want to spend ten minutes cleaning up?

There's not enough space.

The smallest space I have ever seen was the "house part" of a small sailboat that housed a family of five. They were still able to roll out a yoga mat to create a small amount of usable exercise floor space. Yes, you might have so much stuff that you need to move things out of the way when it's time to get down on the ground, but that's movement too.

I have a hard time making myself change my position.

Okay, this one is sort of about your body, but it's more about the inertia everything in the universe has to deal with. All bodies (i.e., collections of matter) tend to continue doing the same thing they're already doing. Getting into a new habit is hard because those old habits have their own momentum, so to speak. It's easier going to the same chair again and again and again. It's not your fault; it's just physics. But this is where "mind over matter" comes in. You're going to have to override the old habit by making yourself do the new *something* (like getting down to the floor). The good news: inertia works both ways. Once you have the habits that serve you better, staying on course is easier because the momentum of those habits helps perpetuate them.

Like I said, carrying your pelvis from standing height to ground level and back up again isn't easy, but it's work that pays off in many different ways, so just do it already. Getting to and from the ground is its own exercise, with its own benefits, but the bigger ticket is that after you become fluid in that motion, you'll be able to do other things— other movements that require the floor—once you're down there. Open your hips, open your experiences.

YOUR POSITION IN LIFE

Almost nobody asks "How much sitting is too much? I want to make sure I'm not hurting myself. Is there a balance to be struck here?" Yet I get lots of versions of a similar question when it comes to standing workstations. How much standing is too much? Is it actually better to stand than sit?

This is a good question, because whether it's sitting or standing, we tend to go looking for the **optimal single position**, as if there aren't many ways you could position your body.

The body positions we use are culturally significant. We tend to think "This is just the way I sit," or "This is just the way I stand." But the positions we use are mostly responses to our environment. Chairs, footwear, clothing, terrain, temperature, gender, class, and fear are a few of the many factors that affect how we position our body when sitting and standing.

When it comes to rethinking our sitting position, what are the options? Turns out there are many. One extremely cool *American Anthropologist* article from 1955, "World Distribution of Certain Postural Habits," reports the findings of physical anthropologist Gordon W. Hewes, who cataloged over a hundred positions used throughout the world. (I love this research so much that I had an illustrator draw a version of the positions that can be used as alternative ways to sit, and now they're on a "Think Outside the Chair" poster that people have hung in classrooms, offices, and homes all over the world!) Most of these positions look similar to stretches and squatting positions we might do for exercise. People in some cultures weave the movement we struggle to find time for right into their daily living activities.

Our perceptions of movement are influenced by how we move. But it also works the other way around. A paper like Hewes's (and hopefully a book like this one) changes the way you think about your body and the way it could move. When we see examples of people all over the world who take a seat, but not always in a chair like we do, our perception can change. In this case, we can think beyond what we consider to be our only two options: sitting in a chair or standing.

Think of sitting and standing as categories, each full of many options, and come up with ten different positions for each. If you have a standing workstation, stand a few different ways every hour. When you sit, sit a few different ways every hour. When you get home, **try sitting on the floor in ten different positions**. Make a note of the shapes your body can't maintain for longer than a couple of minutes and resolve to practice them at the beginning and end of an exercise session. And please don't pass on this cultural practice of sitting in mostly one way. Don't insist that kids sit in the position we did that means we now have to spend time trying to troubleshoot our bodies because of it. Allow them to explore other options. And join them! Kids can teach you something about all the different ways to move.

Hewes concludes his research with this:

> *Physiologists, anatomists, and orthopedists, to say nothing of specialists in physical education, have dealt exhaustively with a few "ideal" postures—principally the fairly rigid attention stance beloved of the drill-master, and student's or stenographer's habits of sitting at desks. The English postural vocabulary is medio-cre—a fact which in itself inhibits our thinking about posture. Quite the opposite is true of the languages of India, where the yoga system has developed an elaborate postural terminology and rationale, perhaps the world's richest. In conclusion I should like to stress the deficiencies in our scientific concern with postural behavior, many of which arise simply from the all too common neglect (by non-anthropologists) of cross-cultural data.*

Addressing this deficiency is pretty much why I do the work that I do. I think there's a relationship between the movements we do, the language that develops to describe them, and the research that investigates questions and concepts formulated from that language. A culture that doesn't move much will lose the ability to move, and it also no longer needs a language that adequately describes the phenomena of movement. Without the movement or the movement language, there's certainly no reason to investigate any questions. It makes sense to me that the science of movement and position is neglected. A thorough science of movement is driven by people moving, not the other way around.

HAVE PSOAS, WILL TRAVEL

Whether by car, plane, train, or foot (she types, looking at her back-packing gear), traveling often involves a lot of hip flexion (found in both sitting and hiking), carrying heavier items than usual, and often that last-minute flurry of added stress. When it comes to my own body, putting these all together sets the stage for old injuries in my lower back and shoulder to flare up.

Back spasms and the like can wreck your plans, so here's some advice from this biomechanist/someone who has had to deal with this issue a few times myself: prioritize multiple short (3–5 minute) exercise sessions every day a week or two beforehand, then do a few exercises en route and again as soon as you arrive at your destination. Think of these moves as travel insurance that's free, but that you still need to put in place ahead of time.

Every time I've had a low-back problem on the road, it was because I jumped right into a bend-and-twist activity (moving luggage, picking up a kid, once it was dragging a vacuum cleaner) without being mindful of the mechanical situation my body had just been through the last few hours (i.e., sitting). **The body needs a transition zone— just a little time to remind it of all the other positions it might have forgotten it could assume after repeating a single shape or pattern for an extended period of time.**

The psoas is a muscle that, when it's able to move well, makes the world glow a little brighter. When it doesn't, it can affect a lot of different parts.

The *psoas major* muscle—you have two of them, one on each side of your spine—is often thought of as a hip flexor. While it can indeed draw the leg bones closer to the torso, unlike other muscles of the leg, the psoas major also attaches to the spine. In fact, each psoas major has attachments to all the vertebrae and intervertebral discs in the lower back.

Because of where it attaches in the body, the psoas has to shorten quite a bit when you sit. When you sit for a long time and when you sit a lot, the psoas doesn't lengthen easily (read: all the way) when you get up out of your chair. Just as this muscle can pull your legs toward the spine, when you stand up and straighten the legs, the psoas can also pull the vertebrae in the lower back toward the legs. So when my psoas is stiff from travel, it quickly can become a pain in my back.

My must-do movement when I'm on the road (or spending a lot of time on the computer writing books about why and how you should get off the computer) is an easygoing low lunge.

LOW LUNGE

1. Start on your knees, adding a folded towel, blanket, or cushion if you need extra cushioning. Step forward with the left leg so your

shin is vertical. You will have close to a 90° bend in both the front and back knees.

2. In order for a lunge to target the muscles that get shortened by sitting, you need to tuck the tailbone under as you do it. This is what moves the muscles' attachments away from each other to create the stretch. Also, make sure your pelvis isn't twisting.

3. Don't let the bottom of the ribcage lift when you lunge. The psoas begins attaching to the vertebra right at the bottom of the ribcage (thoracic vertebra 12, or "T-12"). When you drop into a lunge with a tight psoas, the psoas can pull your spine forward. Look in the mirror as you lunge to see if your torso is leaning

forward or if your ribcage is thrusting. Bring these parts upright so they're stacked right over the pelvis as you've learned to do when standing.

4. Spend a minute here, constantly checking your pelvis and ribcage position, and then do the same thing on the other side, only with your right leg forward. Repeat this cycle three to four times to get that sitting-baggage out of your body.

The great thing about learning which movements your body needs and how to do them is that the knowledge is yours to keep forever. And you can pack it with you wherever you go—it doesn't weigh a thing!

VARUS AND VALGUS

One of my favorite college courses was medical Latin. I loved this class because speaking Latin made me think of gods and goddesses, togas and dolmades, and riding a moped after swimming in the warm Mediterranean. Which, now that I think of it, are all Greek things.

It turns out I love all Greek things and can't wait until I move to my tiny Greek island/olive tree farm/alignment center. You're all invited.

Most people who study anatomy and physiology take a class like medical Latin at some point in their college career. We take it because we have to learn hundreds of names—of the bones and their ridges and crevasses (which have their own names), of muscles and tissues, and of functions. So much of our anatomical language comes from Latin and Greek that learning even a little bit of Latin helps because then the names make sense and you're not just memorizing everything.

Speaking of Latin, *valgus* and *varus* are anatomical terms for shapes of the legs often described more casually as *knock-kneed* and *bow legged*, respectively.

Language is a funny thing, though. Many words have become ambiguous over time, and some are even used to mean the *opposite* of what they used to. Case in point: *valgus* and *varus*.

Valgus is Latin meaning *twisted outward*. *Varus* is Latin for *bent or grown inward*. Originally, the medical terms *valgus* and *varus* referred simply to the **location of the knee joints**. Knee joints that had moved away from the midline were *valgus* and knee joints that dropped inward were *varus*.

Again, the "outward" and "inward" part of the definition referred to the *knee joints* being toward the midline (varus) or away from the midline (valgus).

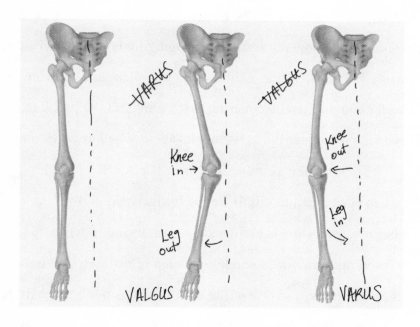

When *varus* and *valgus* were translated for modern orthopedic texts, there was some confusion about which part the "outward" and "inward" labels were describing. Instead of looking at where the knee joints sat relative to the body's midline, the labels were used to describe the direction the lower leg was angled.

In knock-kneed legs, the lower leg angles away from the midline, so now this shape was deemed valgus, or "twisted outward." In bowlegged legs, the lower leg angles inward to the midline, so now this shape became varus, or "twisted toward the midline."

By changing the part of the leg attached to the label, the definitions were effectively reversed. Or maybe it wasn't effective at all.

There's an article in the *New England Journal of Medicine*—"Varus and Valgus, No Wonder They're Confused"—that outlines a problem that began with orthopedic lecturers recognizing how medical students are confused by these terms, chalking it up to them not knowing their Latin well enough, then discovering that Latin isn't the problem. The issue is a lack of consensus in textbooks about whether *varus* and *valgus* refer to the knee joints or the lower leg bones.

There has been a call to eliminate the terms *varus* and *valgus* altogether because everyone is confused (even I, trying to write this essay, got confused multiple times and had to keep referring to my Latin and orthopedic texts, and one of the editors of this book spent fifteen minutes trying to show me I was wrong until she realized I wasn't). *Bowlegged* and *knock-kneed* are clear enough to keep everyone on the same, correct page, even if they don't sound very technical.

I'm not sure this is a good thing, but it does seem like sometimes, if you wait long enough, language can change your position for you.

CHAPTER SEVEN

YOUR ANKLES AND FEET

" You can think of ankle action as an extension of your heart muscle's action. Heart-muscle action has a hard time reaching all the way down to the lower legs. Without your calf-hearts beating, blood becomes more resistant to circulation. Time spent not using your calf pumps is time the heart muscle has to go it alone, which it's not well-equipped to do. THE HEART COMES WITH CALF-HEARTS FOR A REASON."

NOURISH YOUR FEET:
THE BENEFITS OF BEING BAREFOOT

A lot of high-profile nutrition ideas in the past decade or two have focused on some kind of "back to earlier days" approach— eating less processed foods like your great-great-grandparents did, "going paleo," foraging. This idea of finding the intelligence of older approaches isn't limited to your diet, though. You can apply the general idea to the body parts farthest from your mouth: your feet.

Some fast facts: More than 25 percent of the total number of bones and muscles in your body dwell from the ankle down. Each of your feet has thirty-three joints. The intricate design of your feet indicates the potential for them to be about as dexterous as your hands. Really! But imagine someone taping up your hands and how that would affect the way you text and type, catch a Frisbee, or pick things up. You could still do all of these things, but you'd recruit other body parts and might even strain a shoulder along the way. And as for the immobilized hand parts—have you ever had to wear a cast on an arm or leg after a fracture? Casted body parts quickly lose their muscle mass, ranges of motion, and overall performance.

Unfortunately, wearing stiff shoes every day has created a mitten-hand situation in our feet. Not only to the nerves, bones, fascia, ligaments, and muscle, but even to the skin of our feet. In a way, the shoes most of us have been wearing for a lifetime are sort of like a diet filled with ultra-processed foods—conventional shoes are the feet's Standard American Diet. Compared to our feet's natural unshod state, our shoes are stiff, usually prop our heels above our toes, and leave no room for our toes to move. For millennia, we walked on natural surfaces: shifting sand, slippery ice, knotted roots, soft pine needles. Now we hike through the urban jungle—hard, debris-filled surfaces poured over the earth, surfaces that have all been made almost entirely flat and level. Our feet, in shoes and on flat concrete, are sort of like our digestion on a crappy diet—they're starved of the nutritious movement inputs they need for health. And, just as the food you put into your mouth can affect your entire body, your footwear impacts your whole body. We need to transition to being barefoot more, more often, and on more natural surfaces. Some benefits to going barefoot include:

- More foot movement: Once you remove the walls and floor created by shoes, your toes can stretch away from each other, your heels can drop all the way to the ground, and instead of bending only at the ankle and ball of the foot, you can also articulate multiple places in between. Increasing the complexity of the terrain (texture, grade) you walk over also increases the number of movements you (and your feet) experience with each step.

- Increased skin development: Going barefoot doesn't only move the "moving parts" more; contact with the earth moves your skin! Calluses form when the skin is pulled (thicker skin is less likely to tear; your body is wicked smart). The thicker your skin, the tougher your foot armor, i.e., the more resistant it is to thorns and other pokeys.

- Wider range of motion for the ankles, knees, and hips: Shoes impact more than foot parts—the geometry of a shoe, specifically its heel, forces a particular gait pattern. When your heel can't get to the ground, your ankle moves less, which in turn can change the movements at the knees, hips, and low back. Going barefoot is truly a "whole-body" phenomenon.

- More engagement with the world around you: If you're used to always walking quickly, head in the phones, not paying any attention to how you're moving, walking barefoot will surprise you. Nothing pulls you into the moment or has you reading your environment and considering each foot placement like removing your shoes.

Are you convinced and ready to go barefoot? Hold up. There's good news and slower good news. The good news is that by learning more about your feet, you can restore their performance and start to gather some barefoot benefits. As long as your feet contain living tissue (which they do), they can change, grow, and improve, no matter what they've been doing (or not doing) up to this point.

The slower good news is that just as you wouldn't rip off an arm cast and immediately start doing cartwheels, it's unwise to take feet that have adapted to the protection of shoe exo-skeletons and pound them into the ground. You need to transition slowly. Here are some steps to get you going:

1. Create a "no shoes in the house" rule. Besides the occasional Lego, a house is barefoot friendly, so this is a simple way to increase your unshod steps.

2. Practice corrective exercises. This is barefoot time that's set aside to build missing foot muscles. Spread your toes away from each other. Try lifting only your big toes, and then try lifting each toe in order. Stretch your calves, the tops of the feet, and try my Catsuit Stretch to see how tension in the back of the legs can be connected to tension in your lower back and neck (see page 219).

3. Introduce texture. To fully move your foot, you have to step on lumps and bumps. Start with something soft, like a tennis ball, making sure to step on it with all areas of your foot. After a while, create a texture box of pebbles, stones, pinecones, and leaves. Get used to how it feels (maybe icky) and recognize that any aversion might be simply due to a lack of exposure and can change.

4. Take it outside. Start by doing your correctives outside, stepping slowly on the earth in your backyard or local park area,

always checking for debris (and notice how you become more aware and engaged with your surroundings).

5. Take it on the go. Take your shoes off for a portion of your walk (preferably on natural terrain), moving mindfully and taking both uphill and downhill routes for maximum movement. Over time, take your shoes off earlier in your walk until they're not needed at all.

A thoughtful transition toward barefoot time can be grounding, strengthening, mobilizing, and awareness-making. Still, there's no reason to throw all your shoes into the fireplace—you'll still need some form of footwear. However, I strongly suggest you transition your closet to minimal footwear so that when being totally barefoot is not available to you (like, most of the day), you can retain elements of barefootedness—more space for your toes to spread and more potential range of motion for the leg's joints—*with your shoes on*!

Whatever you choose, choose it with a healthy dose of joy and tread well!

EXERCISES TO RETHINK YOUR FOOT POSITION

Calf Stretch: Place the ball of your left foot on the top of a half-foam roller or rolled-up folded towel, drop the heel all the way to the ground, and straighten that knee (image top of page 218). Step forward with your right foot as far as you can without leaning your

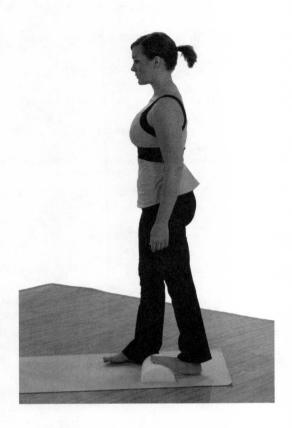

torso forward. Keep your weight stacked vertically over the heel of the stretching leg. Hold for a minute, then switch legs; do this three times each leg.

Top of the Foot Stretch: Holding on to something if necessary, stand on your right foot and reach your left foot back behind you, tucking the toes of your left foot under and placing them on the floor. Hold for up to a minute (stop if you start to cramp, then restart when you can), then switch legs; do this three times each leg.

Catsuit Stretch: Sit on the floor with your feet pressed flat against a wall and your legs straight (sit up on a couple of pillows if your hamstrings are tight). Fold your body forward, tipping your pelvis, ribs, and chest towards your legs. Completely relax your neck, letting your head hang (bottom image, facing page).

LACE UP (OR DOWN) FOR MORE MOVEMENT

Someone once asked for my thoughts on lacing taller hiking boots just to the ankle instead of higher on the shin—a trick she learned from a roller-derby friend.

I loved this question because it recognizes that shoes themselves have different parts that can move and adjust, changing the way they work.

Lacing shoes in different ways is a great way to maximize what you want from a shoe and minimize what you don't. Laces affect how a shoe—and thus your own anatomy and gait—works, and everyone can adjust their laces to serve their body best.

Here are some things to consider when it comes to lacing:

You can adjust shoelace tension (make them tighter or looser) at different parts of the foot to either add or remove support. Search online for "how-tos" on lacing shoes for high arches, narrow feet, etc.

Laces can also prevent or allow ankle motion.

I don't really like taller shoes and boots because I like to keep my ankles free to flex forward and back as much as possible with each step. But the weather and terrain I love often call for the warmth, dryness, and tread I can only find in a shoe shaped more like a boot. At the end of a winter wearing taller shoes, I find that my feet, ankles, and legs are weaker. I need something in between.

I took multiple short walks with my right and left boot laced differently: one to the top and the other down one notch. The one that wasn't laced as high let my ankle joint flex more before my heel had to come up. A heel that stays down longer means the foot is on the ground—*pushing off*—for longer. A longer push-off means my butt muscles are used more with each step, which might be why wearing tall boots makes my hips hurt after a while.

Of course this lacing tip came from a skater! More ankle dorsiflexion with each stride can mean more speed. I described this phenomenon in *Move Your DNA*, using clap skates and speed ice skating as the example. Push-off is important, and to get it, your foot (or skate bottom) has to stay on the ground as long as possible. Anything that reduces dorsiflexion (tight calves! shoes that prevent the ankle from moving!) creates a shorter push-off. P.S. This is also why kids need their ankles free, especially as they're learning to walk and run.

When it comes to how the body moves and feels, footwear matters—and how you lace it matters too.

HOW DO MINIMAL SHOES MOVE?

There are different characteristics that make up a "minimal shoe," and one of those elements is sole flexibility. How do you figure out sole flexibility? With two simple tests.

First, twist the front of the shoe in the opposite direction to the heel of the shoe. How easily does it change shape?

Second test: Fold the shoe in half, bringing the front, toe-holding part of the shoe to the heel-holding part. Does it bend easily into a ring shape?

Why do we care if the sole of the shoe is flexible? In short, there are a lot of bones and joints in the feet. Many of these joints never get a chance to move because stiff shoe soles keep the foot from

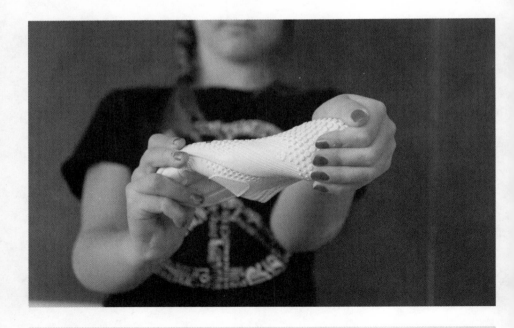

moving (i.e., having to change shape) when walking over something bumpy or textured.

Flexible footwear is usually very thin, but not always. A frequently asked question about minimal footwear goes something like, "What about when I'm standing or walking all day on hard surfaces like linoleum or concrete; how can I find a sole that bends but is still cushiony?"

In this case I'd be looking for a shoe bottom that's both flexible and cushioned. One example of a shoe that meets both these criteria is a Croc-type shoe (nobody said this was going to be pretty). They bend with the feet, but they also protect the feet from so much hardness, which is probably why so many on-foot hospital staff wear them (that and because they wash up really well, which is important because of blood and other fluids).

GAIT 101

When barefoot shoes first entered the mainstream conversation, journalists often asked me to explain the best way to land on feet wearing minimal shoes. I would ask for clarification—for walking or running? To which they'd reply there wasn't room in the article for the difference. Couldn't I just quickly tell them how to use barefoot shoes correctly? In five sentences?

Once you've known me long enough, you'll know I have never been able to clarify anything in five sentences. And while the mainstream frenzy over barefoot running has come and gone, the fact remains that minimal shoes are excellent tools, and our gait in them and barefoot matters a lot. It's essential to understand the differences between running and walking gaits, so let's get to work.

WHAT IS GAIT?

"Gait" is a term that refers to the *way* a person walks or runs. It's the repeated pattern of arm and leg movement that occurs to move you over ground on foot. Just as everyone uses their personal speech pattern when they talk, people have a particular gait pattern when they walk.

Our gait patterns are the expression of many influences: learning to walk as a child by mimicking those around us, our lifetime of movement habits, footwear, any injuries occurring along the way, and straight up any FLAIR that we just add in.

Our gait patterns are so specific to us that there is now gait-recognition technology—software that links a person's gait "fingerprint" to their identity so they're easy to spot from video surveillance footage. So it turns out the reason "you can run, but you can't hide" *is because of the way you were running*.

We have both a walking gait and a running gait, and these are different because walking and running are different movements (running is not just a faster version of walking). Biomechanists and physical therapists are concerned with gait patterns because the *way* we move (and not just whether we're moving at all) creates the loads experienced by the various parts of the body. Our tissues are influenced by the loads we create, sometimes for the better and sometimes for the worse. Thus, we are on the lookout for abnormal gait patterns—missing or unnecessary movements of a gait cycle—that influence body injury, decrease balance, or create other undesirable scenarios.

Walking and running differ in a few ways, but a primary difference is that running has a flight phase during which *no feet touch the ground*, compared to walking, during which there's always at least one foot on the ground.

This flight phase creates another difference between walking and running: the higher loads running creates.

A QUICK LESSON ON LOADS

When you stand, you only need to deal with your body weight. If you weigh 180 pounds, then your body needs to deal with 180 pounds of force when standing.

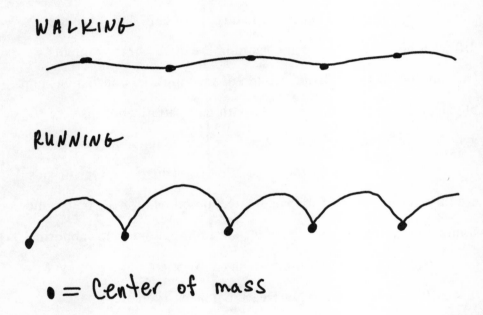

WALKING

RUNNING

• = Center of mass

During normal walking, body weight is mostly traveling straight forward. The body's center of mass moves up and down a tiny bit, so you create heavier loads compared to just standing. Walking creates loads of 1.5–1.75 times your body weight. If you weigh 180 pounds, then your body needs to control 270–315 pounds while walking.

Running creates loads 2–3 times your body weight. This is because instead of your body staying level as it travels more or less forward, your body (its center of mass represented by the dot on the image below) repeatedly moves up into the air then back down to the ground. Running is like a series of mini-leaps. The body of a 180-pound runner deals with 360–540 pounds of force. (Jumping creates even higher loads than running, but in the case of jumping, you're usually just doing a few in a row, not going for miles.)

Have you ever jumped and landed with straight knees? It feels pretty bad because it jars the entire body. This is why we do things like bend the knees and land on the balls of the feet when we jump— and we often do slightly less dramatic versions of these when we run and walk, too.

Athletic shoes have lots of padding under the heels, and landing on that squish reduces the impact—or so the story goes. When you don't have cushy-heeled shoes (like most humans until recently) and have grown up running barefoot or in thin soles, the "cushion" is created by a particular gait pattern: landing on the front of the foot and using the calf muscles to quickly slow down the descent of earth-bound heels so they don't slam into the ground.

When you're *walking* barefoot, you don't automatically need to land on the front of the foot, because the loads are much lower. Because you've always got one or two legs on the ground, the muscles down the outside of the legs (the lateral hip muscles) can efficiently carry the load in a way that keeps your center of mass moving forward smoothly. However, when we walk barefoot and are unsure of what's under us, we often tend to land on the front of the foot to keep things as light as possible in case we land on something sharp or wobbly.

Just because there's an efficient way to walk or run doesn't mean everyone does it that way, though. Like I said before, many factors contribute to our walking and running styles. Those styles are just habits, and the cool thing is that, like all habits, they can change. We can all learn to adjust the way we walk and thus the way our body is being loaded and shaped with each step.

GAIT 102

I've always been able to see what people's joints are doing when they move, even when they have their clothes on. I don't know how or why, but movement pattern recognition comes naturally to me. I was hoping it would be something like writing operas or painting frescoes, but that's not what I got. What I got plays out more like this:

Katy (watching the movie *Superman*): That guy playing Clark Kent as a kid is limping.

Husband: Shhhhh.

Katy: There's something wrong with his hip.

Husband: Shhhhh (but reaches for cell phone).

Katy: I love Superman.

Husband: Who doesn't?

Katy: Shhhh.

Husband: The IMDb website says that the kid tore his hamstring muscle while filming. That's why he's limping.

Ha! Katy *1*, Superman *0*.

Like the baseline movement of many animals, human walking and running have evolved to be efficient movements created by the complex coordination of many lever systems. Gait inefficiencies are any movements that make walking or running less efficient (read: harder on the body). But before I go further, it's important to say that context matters when it comes to understanding "normal" gait. We come from a sedentary (think driving vs. walking), shoe-wearing culture—an elevated-heel shoe-wearing culture at that—that has done most gait research with young athletes inside on treadmills.

But in real life, walking (and even running) is more like a broad category of movements. Walking on a treadmill in a lab is different from walking on uneven terrain with unknown textures underfoot. If you're barefoot and walking at night through an area you don't know, your gait pattern will instinctively differ to when you're walking through a clear, familiar area in full daylight. There are uphill and downhill walking patterns, patterns when you're carrying things like water or children, and the way you walk when it's icy underfoot. Each of these scenarios will result in a gait that doesn't look the same as what you would find in a lab, but that doesn't make it inefficient. Also important: although gait differs depending on the context, we can still optimize how we move through different scenarios.

That all being said, it's easiest to work on improving plain old baseline gait on the flat, level, familiar surfaces we walk on every day. Around your neighborhood, aisles of the grocery store, airport, school halls. Just regular old walking is a great place to begin observing.

What I look for are things like torsos that lean right and left with each step instead of staying vertical; knees that bend a lot with each walking step (behaving more like running knees); pelves that twist with each step, pelves that tilt forward and backward in time with the leg (indicating a lack of hip mobility), and heads that bob along— again, seeing elements of running in a walking gait. These little movements are easier to spot in others than they are in ourselves. It's hard to see yourself walking! When you're first starting to work on your walk, have someone film a short video of your walk from the front, back, and side so can see yourself more objectively.

Our gait inefficiencies serve us in some way. They enable us to move given the current state of our body. For example, a torso that rocks to the right and left does so for a reason: it puts less weight on the leg when it goes to step. Perhaps the body-rocker has a painful knee or hip, or weak legs. It makes walking easier and more efficient for the body at the moment, and that is the case with most gait inefficiencies.

So what's the problem? We need to be concerned with the longer game. Compensatory movement strategies that work in the moment can contribute to injury in the future or allow degeneration to persist and progress. They're inefficient in the long term and can lead to significant pain and reduced movement. I think we need to celebrate what we can do and what we've been able to do while also acknowledging that we can improve. Investing energy in learning about and changing our own movement patterns for the better is worth it.

GIVE YOUR HEART A LEG UP

I'm tired of looking for recipes online only to find them buried at the end of trillion-word essays, so I'm going to start you off with this exercise at the top.

CALF ELEVATORS

Calf Elevators are similar to calf raises; the main difference is you work extra hard to keep the ankles from wobbling. (Ankles, like elevators, should be able to move the body straight up and down without traveling side to side or rotating as they do so.)

1. Stand with feet forward, ankles pelvis-width apart.
2. Slowly lift and lower the heels, keeping your ankle joints centered and moving straight up and down (don't let them drop outward or inward, or twist). Don't overly thrust the pelvis to help you get up, and keep your toes lift-able throughout the entire exercise (weight is on the front of the FOOT, not the toes).
3. Want to level up? Start this move with your toes placed slightly uphill on a folded towel or book. This will give a greater range of motion over which your calf muscles have to move!

Now that you've moved a bit, are you ready to blow your mind a little? Consider this: Moving your ankles is one way to take care of your heart.

Your heart muscle moves constantly, contracting to pump and circulate blood throughout the body. Oxygenated blood flows *away from the heart* via the artery-parts of the cardiovascular system, and deoxygenated blood moves *back to the heart* via the venous-parts of the cardiovascular system.

Blood moving away from the heart is under more pressure, so it all flows forward, but the blood moving back toward the heart is not only under less pressure, it often has to move uphill to get to the heart. This means returning blood has the tendency to flow in the wrong direction.

Thankfully many veins, especially those in the arms and legs, have vein valves—flaps that shut behind the returning blood—keeping the blood moving in the right direction. You can think of venous blood like a salmon moving "uphill" through the limbs' veins and the valves as a "salmon ladder" the blood uses to climb.

The valves are essential, but they're also passive. They don't help the blood move forward, they just stop it moving backward. There is, however, an active way our body can assist venous return—a more powerful phenomenon that you may or may not be taking advantage of: the calf muscle pump.

Veins are often embedded within muscle. When the calf

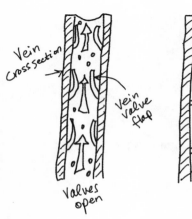

Vein
cross section

Vein
valve
flap

Valves
open

Valves
closed

muscles repeatedly contract—*short, long, short, long*—it surrounds the veins with a *tense-relax-tense-relax* pumping action that propels the blood up the legs. In other words, calves are like tiny extra hearts stored at the bottom of your body that work when you want them to. They are CALF HEARTS!!!—and they have an important role in the all-day action of moving blood through your body. And to be clear, we're talking here about the calves of your legs, not baby cows.

You can figure out your heart's beats-per-minute by counting your pulse for sixty seconds, but how do you figure out how often your calf-hearts are assisting? You can measure calf-heart heartbeats in steps per day, or the number of times your ankle moves through its plantar-dorsiflexion range of motion. How many calf-heart heartbeats do you create in a day?

We know being sedentary is hard on the heart, but it's not always clear *why*. It's no wonder; it's complex! In this case, you can think of ankle action as an extension of your heart muscle's action. Heart-muscle action has a hard time reaching all the way down to the lower legs. Without your calf-hearts beating, blood becomes more resistant to circulation. Time spent not using your calf pumps is time the heart muscle has to go it alone, which it's not well-equipped to do. THE HEART COMES WITH CALF-HEARTS FOR A REASON.

INVERTED LEGS VS. STEPPING LEGS

When vein valve flaps are under lots of pressure and are not supported well through calf-muscle contraction, they break. The result is varicose veins. One of the movements advised in this case is to lie on your back with your legs up a wall. As long as you're in this inverted position, the blood can move back toward the heart with gravity's assistance. But how much time can we really spend upside down? The human body has been an upright mover for some time, so how can you make it easier on your cardiovascular system for the bulk of the day? Get your calf-muscle pumps going with regularity.

Other ways to get your calf-hearts pumping:

- Calf Stretch (page 217) to increase the ankles' range of motion and therefore their capacity for pumping action.
- Take two or three fifteen-minute walks each day and really focus on pushing off at the ankle.
- If you can't walk, lie in bed and point-flex-point-flex your feet for a few minutes every hour.

Strong, healthy feet matter to more parts than just your ankles, knees, and hips. Improving their movement matters to your heart!

ABOUT BUNIONS

Hallux is the Latin word for the big toe. It was derived from the Greek verb "I spring or leap," which makes sense because the big toe is very much like the pole used by a pole vaulter. The tip of the big toe is the last part of your body to leave the ground after your foot (and all that's stacked on top of it) vaults over into your next step.

Hallux Valgus is the term for a big toe that points laterally (toward the pinkie toe) instead of forward. Over time, the big toe can be pulled so far toward the pinkie toe that it becomes almost perpendicular to the other toes. After months and years of walking—*vaulting*—over a toe in a sideways position, it can start to hurt. How does the body deal with loads on a sideways toe? It bulks the area up into what we call a bunion.

The term *bunion* also comes from the Greek language, meaning small hill, mound, or heap. The growth of the bunion is a response to *excessive loading* occurring at the joint of the hallux. Bunions are created by this inappropriate loading (using the joint in the wrong position).

People are commonly told that bunions are genetic. It's true that some people have genes that cause their collagen content to be proportioned differently, meaning their toes are more likely to be unstable and go sideways more easily, but the percentages of bunions caused by this gene variant are very low (less than 10 percent). Anyone with this collagen issue will experience it in multiple tissues and joints. If you don't have a whole-body collagen problem, any bunions you have are most likely caused by habits. Even if you do have a collagen issue, you can often improve the stability of your joints (including your big toes) through really mindful corrective exercises.

Bunions are certainly more common than collagen issues. About one in four adults have them, or one in three in people aged over sixty-five. They are twice as common in women. They can *seem* genetic because your parents (probably mom) had them, as did her mom, and her mom's mom, which seems like a lot of moms. With so many moms, surely bunions must be another trait passed down to you via shared DNA, like amber-colored eyes.

So are your bunions handed down? Well, even if you didn't inherit bunion-allowing collagen from your parents, it's entirely possible you picked up similar gait patterns and adornment habits from them, just as you might share the same accented speech. So while bunions aren't usually a genetic inheritance, they can be an inheritance nonetheless.

Does your family line have footwear in common? A lot of us grew up wearing shoes that pressed our toes together too tightly. "Good" shoes. "Proper" shoes. Our big toes have been pushed toward our pinkie toes for many years. The muscles between the toes tighten, reinforcing this sideways push. Some shoes, especially fashionable high heels, not only push the big toes towards the other toes, they also place the weight of the whole body right smack over the big-toe joints. Now your weight is right on top of the sideways toe. Wear shoes like this for a night and you can feel soreness in your feet. Do this over a lifetime and the hallux starts to beef itself up (i.e., make a bunion), the joint starts to stiffen, and walking becomes painful.

NEW HABITS

The good news is that while you can't ditch the amber eyes, you can change your "inherited" habitual positions and add in corrective exercise. Your muscles and connective tissue are quite pliable and respond to changes. When your feet are all squished together, the adductor muscles between the toes get very tight, but they are easily stretched out. Begin with a daily stretch of all toes, giving extra attention to the big guy. Place your fingers between all of the toes to help restore adductor muscle length and joint range of motion. If you have a pretty significant angle on the hallux, spend some time gently moving it away from the toes by itself.

Let's talk about gait changes. Bunions aren't only made by shoes, so even if you have great footwear and don't need that much stretching, you might be able to reposition your feet for huge benefits.

The "vaulting" action of the foot is maximized when the toe and foot levers point forward. Many of us have developed (or, you know, "inherited") a slight to exaggerated turnout of the feet. We still walk forward, of course, but we're repeatedly shoving the vaulter's pole off to the side. The simple act of stepping with our feet turned out is part of what keeps pushing the big toe out of alignment.

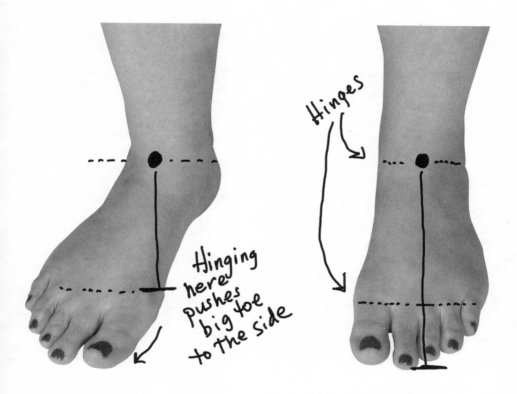

Check your turnout: Use the straight edge on a rug, a tile seam, or the spine of a book. Line up the outer edges of both feet along a straight edge to get a sense of "straight forward." The toes may point in a bit when you first start.

Feel pigeon-toed? That's normal in the beginning. But it's a crucial change as far as the big toe is concerned. When the foot is straight, the big toe won't continue to be pushed out of the way with each step. Of course, if the big toe has gotten to the point where it's too stiff to extend when walking, then walking with your feet forward won't work. In this case, just start 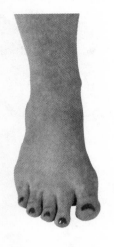 turning your feet as far toward straight as you can, so your big toe can increase its motion a little bit at a time. Supplement with lots of toe stretching, big toe lifts, and toe separators that can stretch muscles between the toes—especially between the big and second toes.

Here's your new-toe-do list:

1. Walk with feet pointed forward when walking
2. Using your fingers, stretch all your toes away from each other so they spread wide
3. Big toe lifts
4. In bare feet, practice lifting and lowering the big toe *by itself* (don't let the other toes lift too).

This information might not be pretty like amber eyes, but feel free to pass it down (and back up!) the line to your loved ones.

STRAIGHTEN YOUR FEET

Like tires on a car as you're moving forward, human feet give their best leverage when pointing forward. In this position the ankle's levers and pulleys can work best, the muscles of the feet and hip can create the foot's arch-shape, and the toes can move more freely.

But feet aren't shaped like tires, so which part of the foot do you line up to check where it's pointing? I recommend using points on the outside of the foot and not any of the toes, and here's why: toes can sneakily go askew. They quickly get pushed around by forces created through shoes and gait.

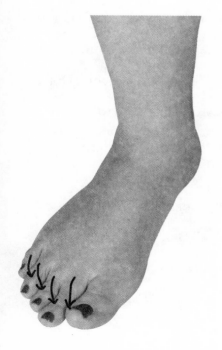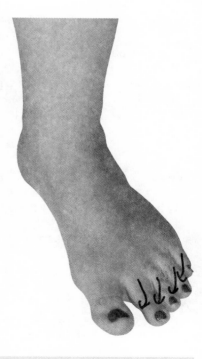

I think of toes as tiny baseballs, your leg as a pitcher, and every step a pitch. For example, if your feet turn out when your leg swings forward while walking, the forward momentum of the leg throws the toes forward.

After years of stepping this way, the toes are trained forward relative to a turned-out foot (image previous page). When you position your feet straight forward, your toes now pull toward the midline. See how it's sneaky?

This is why I don't use the second toe—a common alignment cue used in mat-based exercise classes—to determine foot position, because it works the other way around: foot position determines toe position over time.

To align the foot, I prefer to use points on the actual foot. And P.S. Even feet change shape when you start exercising them more, so you have to be picky and choose parts that are the least malleable. Don't use the toes! Toes are like teenagers. They do what they want.

WHAT'S THE POINT OF FEET?

When you first start lining up your feet, using a straight edge is a simple way to get a sense of "feet forward." But, feet come in many shapes, and not all of them have a straight outside edge. For this reason, we like to take the next step and use POINTS on the foot.

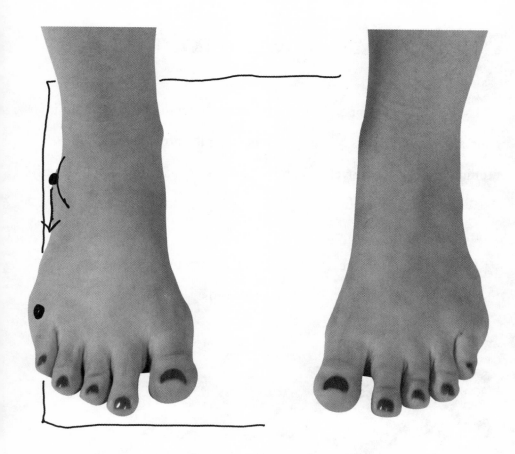

Step on a piece of paper and look down. You'll be using the paper's edge to get a point on the front and the back of the foot on the same line, which will give you a different feel for "feet forward." Position the front of the foot by lining up the bony prominence just below the right pinkie toe on the paper's edge, as in the image above. Position the rear of the foot so the right ankle's lateral malleoli (the round, coin-shaped bone on the outside of the right ankle) sits just over the paper's edge.

THE HIGHS AND LOWS OF FOOT ARCHES

Here's something about me: I love animal tracking and prints. Animal signs are visual traces of past movement, so it's no wonder I enjoy tuning in to all the movement around me—I've made a life of it, clearly!

While I am tickled by wild animal tracks, my favorite animal tracks to read are *human* footprints.

I snapped this picture of my husband's footprints the other day because a picture is worth a thousand words.

In general, there's more talk about "flat feet" and what to do about them in the minimal footwear community than there is regarding the other end of the foot-arch spectrum: "high arches." As you can see in the picture to the left, this guy's arches are so high that the part of the foot that connects the front to the back barely touches the ground.

Feet come in a wide variety of shapes, which are often classified by how they look or measure. One of these measures is "arches," which are then classified as high, low, or neutral (somewhere in the middle). To illustrate this, I made all the people at my house paint their feet and then walk all over some butcher paper, which is probably why nobody ever wants to come over.

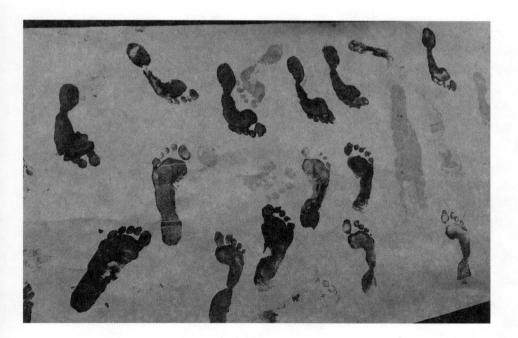

The feet of two adults, four kids, and a dog are captured here. Do any high-arch prints match those pictured to the left?

The photo below shows the highest and lowest arches in
our family.

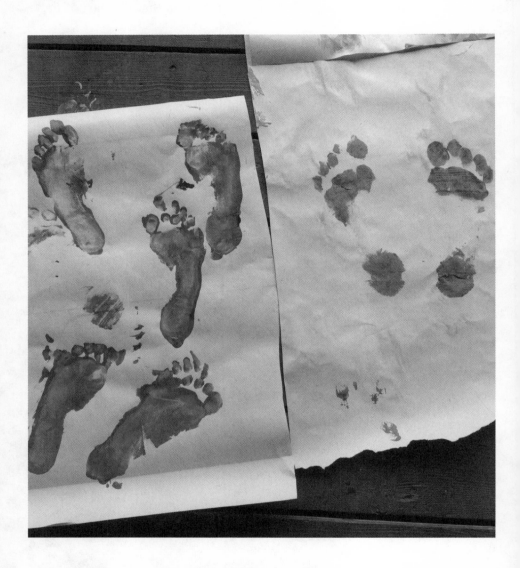

You already know who has the high ones, but the low ones belong
to an almost ten-year-old. (When he was a baby he'd sleep next to me,
both of his feet cradled in my hands, and now they are as big as mine

and I can no longer hold them at night, but he still thinks he's a lap dog and likes to cuddle at night so I'm not complaining.)

Another thing about animal prints of all types: the substrate (what's being walked through) matters. The shape of some of these footprints left behind changed a bit based on how much paint or mud people were slipping on, as well as where their bodyweight was on their feet (and other things too, like how fast or in what direction they're moving). A standing foot (like the right, high-arched foot above) has almost no lateral footprint, but in the moving prints on page 244 on the wood, you can see where the right lateral foot touched down. This could be because we're heavier while walking than standing, so that extra weight pushes the foot down a bit, or it could be because the body (or ankle) leaned to the right a bit due to the direction of walking.

I'm just saying that what starts as something seemingly clear or obvious quickly becomes more complicated the more you learn.

Now about the arches, especially high ones, and what to do about them.

High arches, like really high ones, can make for stiff feet, which means the ankles end up bearing quite a load. They can also make for ankles that hurt and are easily injured. To take care of his ankles, my husband has to work hard at making his feet more supple.

When the arches are even a little more flexible and his lower legs stronger, he has better support for his ankles and the connections around them.

His program specifically:

- lots of Calf Stretch (page 217), Strap Stretch (page 182), and massaging the soles of his feet on a ball
- minimal footwear (especially thin soles)
- taking the non-flat/textured path as much as possible (the more the ground can rise to meet his arch, the better it feels; high-arched feet on flat and hard makes for less support, which is probably where the concept of filling in the arch with a shoe comes from)
- lots of barefoot time (in smaller doses, in safe areas at first)

The stiffness in *your* feet might look different.

Maybe you need to find your arches again. Surprisingly, the moves to help with that aren't that different from the ones to mobilize high arches. Flat feet need to soften (roll the soles!) in certain areas in order to be able to lift at the arch. You'll definitely need some Pelvic List (page 107). We all need certain moves for our bodies, but what we need to supplement with depends on the bodies and movement diets we bring to the table. We just might need less of some, more of another. All our feet could benefit from more movement and having arches that can adjust depending on how we're using them.

And this isn't only a lesson for you to read about. Take a good look at your footprints—as I did ours, by using paint and paper or by stamping and walking wet feet on concrete.

What do your footprints look like now? And, what will they look like after a few months (or years) of regular exercise to make them stronger and more mobile? Footprint record keeping…NOT ONLY FOR NEWBORNS!

STEERING YOUR FOOT LOAD

I was recently featured in an article on foot pain, along with a couple of podiatrists. This line caught my eye, and I felt it could use additional nuance:

"For every pound we weigh, 1.5 pounds of pressure is exerted on the foot, says Dr. Tumen. So for a 175-pound woman, every step feels like 263 pounds of pressure."

Clarifying loads: When you're standing, your feet feel just your weight (so, 175 pounds in this example). Walking loads the feet with 1.5 times your body weight (as mentioned above), and when running, loads are more like 2–3 times your body weight, so 300–500ish pounds in this example.

Clarifying pressure: "Your feet" aren't feeling that pressure as much as CERTAIN AREAS on your feet are feeling that pressure. It's an important clarification, because you can control, moment to moment, where the pressure resides in your foot by playing with your posture and footwear. Your center of pressure is malleable and dynamic. Changing bodyweight can take a long time, but you can change alignment in an instant and immediately alter where the pressure falls.

There are multiple moving parts in your feet, and each affects where the pressure occurs within the foot. Try starting with this simple body-alignment adjustment to see how it feels:

Back your pelvis up to take the weight off the front of the feet, and share your body weight evenly between both feet. This is tricky to do in heeled shoes (they are part of what pushes your weight forward, creating pressure on the front of the foot), so practice unshod at first.

My point, as always: you have more control over your physical experience than you might realize. You're the one steering the load.

PLANTAR FASCIITIS AND TIGHT HAMSTRINGS

I'm not a subscriber to *Foot & Ankle Specialist* magazine, but I have awesome friends who are on the lookout for articles that I might like, and my podiatrist friend sent me an issue's "Clinical Research" section featuring research looking at the relationship between plantar fasciitis and hamstring tension.

P.S. Everyone should get a podiatrist friend. Mine is awesome, because not only is she an excellent doctor, she is also really good at cribbage. Almost as good as me. And if I ever get a cribbage injury, she'll probably treat me for free. Although she won't because she's not allowed to treat anything but the feet. So if I get a cribbage injury in the lower leg—say, I accidentally step on a cribbage peg—I'll surely be saving some money. I love my friends.

Anyway.

One of the most common ailments to show up at the foot doctor's office is plantar fasciitis. The *plantar fascia* is the band of connective tissue that runs underneath your foot along the arch, and *-itis* means it's inflamed. Usually it's not the whole plantar fascia that's inflamed, just the area where it inserts into the heel bone. For most people plantar fasciitis shows up as "the underside of my heel hurts really badly."

To help you visualize how things connect, the foot structure (sort of) looks like this:

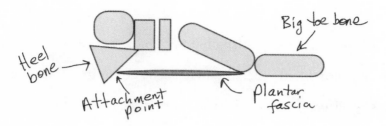

I was excited about the *Foot & Ankle Specialist* article because it described the first study I'd seen of a link between tension in the hamstrings (the large muscles down the back of the thighs) and plantar fasciitis.

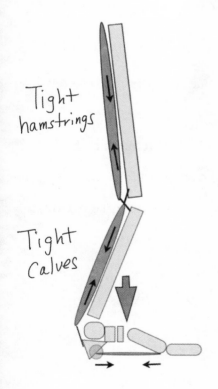

It got me thinking about how hamstring tension could relate to increased tension in the plantar fascia (think "extra pulling where it inserts"), and this is what I came up with: tightness in the hamstrings (like tension in the calf muscles) can chronically shift body weight forward and right on top of the plantar fascia.

This is another crappy diagram, but hopefully it does its job, which is to show how tight hamstrings can cause the knees to bend slightly, increasing the load on the midfoot which places the entire plantar fascia under tension.

Make sense?

Plantar fasciitis is often attributed to tight calves, which is why calf stretching is commonly recommended. The condition is also believed to be a side effect of being overweight (the idea being that too much weight on the foot tissue strains it). This paper provides more nuance by showing that in bodies with a body mass index under 35, tight hamstrings correlate to plantar fasciitis more than body weight does.

Learn to take the tension off your midfoot (and thus the plantar fascia) by simply adjusting your standing and walking alignment. Here are three ways:

- Back up your hips so the weight of your pelvis rests over your heels, not your midfoot or toes.
- Lower your shoe heel height to ground level (remember, heeled shoes move your weight forward and over the midfoot).
- Do exercises that stretch tight calves and hamstrings so backing weight off the midfoot is easier.

ROAD TRIP ALIGNMENT

Long road trips can be a pain. In the back, in the legs, and in the feet. For the same reason you'd optimize your car's wheel alignment before driving, you can also optimize your body alignment while driving.

Next time you drive, pay attention to what your foot is doing while on the accelerator. Yes, the ankle needs to plantarflex (foot pushes down away from the shin), but what you might find is you're not only pushing the foot down, you're also curling your toes. Practice relaxing the toes as you push the accelerator. You can even try toe spreading while pushing the gas. In either case, you're separating calf movement from toe movement. Curling the toes doesn't make the car go, yo.

Another common driving alignment tip for those with low back or sacroiliac joint issues: watch out for the one-legged reach. This is less of an issue in a car with a manual transmission that uses both legs, but in automatics, your right leg is doing most of the work. It also spends time straight when the left one is bent. The pelvis slightly twists to help out. Think of all those hours you're spending in a position that has you slightly twisted to the left from the lower back down!

But guess what? Car manufacturers know that your hips should be ergonomically balanced at the wheel, thus you'll often find a "balancing pedal" on the left side of the floorboard. That tiny little

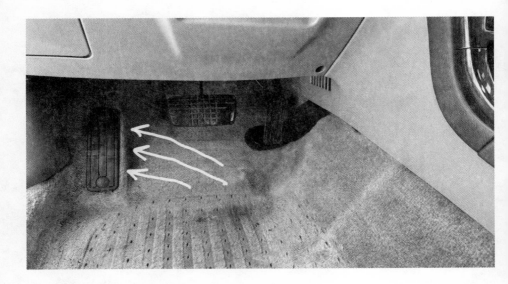

pedal that does nothing? It's where you rest your left foot to help keep your lower body alignment more even.

Keeping your hips even requires a lot of attention. You'll quickly drop right back into your old habits if you stop paying attention, or maybe that's just me. Just as you do with any meditation, start again, start again, start again. You will eventually arrive.

CHAPTER EIGHT

ALIGN YOUR MIND (AND BRAIN)

> " Moving more also helps maintain your brain mass. Just as muscle mass tends to decline with age, parts of our brains are apt to get smaller as we grow older. Physical activity helps maintain muscle mass and also helps reduce shrinkage of hippocampal volume in older adults— especially in those with a genetic predisposition to Alzheimer's disease."

YOUR BRAIN ON MOVEMENT

Until recently scientists believed that the brain received a steady flow of blood (called *cerebral blood flow*), no matter the body's behavior. The brain is always getting enough blood to keep us alive, but better measuring tools are revealing that skeletal muscles aren't the only body parts getting more blood when we get moving; the brain gets more blood too.

In 2017 I sat on the panel of a Women's Alzheimer's Movement event in California hosted by the organization's founder, Maria Shriver. One big takeaway from all experts was the importance of movement to brain health. It's not clear why exercise is protective, but two of the emerging risk factors for Alzheimer's disease are brain *hypoperfusion* (too little blood flow to the brain) and arterial stiffness ("hardening of the arteries")—both of which can be reduced through exercise.

Getting any exercise is better than none, but because the reasons movement is good for the brain aren't clear, we are still missing details about what type, how much, or for whom exercise works best. But there are things we do know, and here are a few relevant tidbits I shared on the panel because I think they're fascinating and because they add nuance to discussions of movement that I think is often missing. We hear "movement is good for you" without specifics that help us rationalize why we need to prioritize it over something else we deem as good.

Get your feet on the ground! More blood gets to the brain when we exercise, but how, exactly? Can it be any type of exercise? At least one of the ways that moving gets more blood to the brain is via the impact certain movements create. When the foot pushes against a walking, running, or cycling surface, a pressure wave is created that regulates blood supply to the brain. The more impact, the bigger the wave, so running (highest impact) creates a bigger wave than walking does, and walking creates a bigger wave than cycling, during which there's very little impact. If you currently take all of your exercise seated, lying down, standing in place, or on a machine that holds your feet in place, mix it up by adding activities that get you rhythmically "pounding the pavement."

Chew your food. It's not only your arms, legs, and abs that need exercise. Two of the strongest muscles in our body are our jaw muscles (masseters), and most of us have undermoved jaws. Although these muscles are relatively small, they can exert the most pressure of all the skeletal muscles. Modern diets leave them pretty sedentary, and that can be a problem for the brain. It turns out the muscular action of chewing helps preserve the health of the brain part (the hippocampus) that deals with memory and other cognitive functions.

I've said it before and I'll say it again: the job of each muscle isn't only to move the levers they are attached to. Muscle contractions have other functions that have nothing to do with moving us around.

Why does chewing support brain function? Perhaps for a couple of reasons. For one, chewing increases blood flow to the brain. The harder you chew—i.e., the more strongly you contract your jaw muscles—the more blood flow to the brain, which researchers measured by having participants chew gums ranging from soft to super-hard. It could also be that brain parts involved in eating are stimulated by the mechanical process of chewing. Or perhaps it's something else. Regardless, there is evidence that chewing preserves brain function, so take a look at your diet to see how often and how hard your jaw works on a daily basis. Is your jaw doing the equivalent of sitting in a chair all day?

There are now jaw exercisers on the market—rubber squares to bite down on so you can do your "jaw push-ups" and "jaw lunges." But what about non-exercise jaw movements? Those movements of the jaw that could fit into each day, no tool required? How many of your calories do you drink versus chew? How soft is your food? Sure, green smoothies are handy and full of dietary nutrients, but what about the mechanical nutrients? You can likely find the jaw movements you need by adding more texture (raw apples and whole carrots, nuts, jerky, and dehydrated fruit) to your diet simply for the exercise of it. Vitamin texture. NOT ONLY FOR FEET!

P.S. If chewing different foods (ripping and tearing movements with our front teeth, breaking down bigger bits with the back teeth) offers a variety of movements and thus jaw loads, the repetitive movements of gum-chewing is like taking your jaw on a long walk. And when I was a kid and shoved an entire packet of Big League Chew gum into my face over a couple of hours, this must have been akin to walking fast up a steep uphill, because my jaw muscles were burning. My hippocampus must have been RIPPED when I was nine.

Move for more brain mass. Moving more also helps maintain your brain mass. Just as muscle mass tends to decline with age, parts of our brains are apt to get smaller as we grow older. Physical activity helps maintain muscle mass and also helps reduce shrinkage of hippocampal volume in older adults—especially in those with a genetic predisposition to Alzheimer's disease. In general, more physical activity is associated with a larger brain volume as one ages, but as with chewing, the precise mechanism isn't yet clear. Whatever the reason, "move it or lose it" applies. That may sound harsh, but my intention is to do the opposite of harm.

You can move outside the exercise box. Humans have been doing many movements in their regular daily life for eons, many of them *aerobic* (increasing your heart and breath rate) without being that thing called "aerobics"—that is, exercise bouts specifically designed to move your body rhythmically for an hour to reap the health benefits. The concept of "exercise" has only recently become synonymous with "movement," and that's only happened in societies that have eliminated physical activity from almost every domain of life. Now it's the only kind of movement we tend to consider valid. Conventional modes of exercise are certainly beneficial, but your brain can benefit from *many* types of physical activity, not just exercise. Gardening, working on a home project or chores that have you moving in new ways, walking or biking instead of driving ("active commuting"), or dancing are all brain-healthy activities and might be easier for you to fit in multiple times each day.

It's true that we don't yet have clear answers as to why movement is essential to the brain, and nor do we know the optimal "dosage" of movement for each individual. But moving more has so many benefits for all your other body parts and so few (if any) negative side effects that it's always a good use of your time. One reason I spend so much time teaching people to focus on how they're moving their individual parts is because it's often a single hurting part that keeps you from moving your entire body. Most of our anatomy and physiology, including the brain, depends in some way on our whole body moving regularly.

It's never too late to begin some brain-healthy movement habits, so maybe right now is a good time to put down this book and feed your brain some movement. Go on, I'll be here when you're done.

YOUR MIND ON MOVEMENT

I once heard a psychiatrist say that at the core of our experiences are the three relationships we are in: with our body, with others, and with ourselves.

Quite startlingly, modern life sets us up to barely meet our bodies' biological needs: for water, nourishing food, rest, and movement. For those of you reading this book, it's likely you barely need to move to get anything you consume. That means movement is now something we have to intentionally seek out.

In recent decades, the act of moving on purpose—especially exercise-type movement—has been tied up with the idea that it improves how you look. This can get jumbled up with how we relate to and communicate with ourselves, that we're unsatisfactory, never good enough, and worse. Physical activity has well-established mental health benefits, yet we continue to focus on how it affects our physical appearance rather than how our bodies function and how we feel because of how our bodies function. We focus less on the simple idea that our *body has a need to be moved, and the way we choose to move our body is part of our relationship with our body*. Outside of very rare cases, you are the only one who can meet your body's basic need for movement. Whether you meet it or not influences how you process the world.

When we don't feel good and can connect that with the state of our mental health, we still collectively struggle to take the next step: connecting some of that feeling to the way we care for our body. Many things impact our mental health, including our genetics, but it's still always helpful to ask your body, "Which biological need of yours could I be meeting more regularly?"

When my children were young, they didn't understand that their feelings—especially when they went reeling or reacted strongly to something that didn't usually bother them—could often be tied to fatigue, hunger, too much sensory stimulation, or too little sensory stimulation. Through language and experimentation ("Eat this and let's wait ten minutes to see if you feel the same way," or "Let's take a rest and see if this continues to be frustrating"), they—and who am I kidding, me, still, today—continue to learn that often the way they're mentally connecting with the world is influenced by the body asking for something it requires but is not getting.

We understand this phenomenon well in relation to food. Who hasn't felt emotionally altered after a huge dose of sugar or gotten hangry now and then? But there's less understanding of our hunger for movement and how it can have just as drastic an effect on our mental wellbeing.

Learning to communicate with your body takes time. The signals the body uses are primitive and not specific. For example, it's still hard for me to remember that feeling cranky and as though everyone around me is super-annoying or hurtful almost always relates to the fact that I've been sitting inside most of the day. Remembering (or being gently reminded or even invited by a loved one) to take a quick walk around the block almost always makes those feelings evaporate. Is everyone getting together to decide to stop bugging me so much once I get back from my walk? Not sure. Either way, I've learned to reframe "everyone and everything is bothering me" as "this is how my body tells me I need to move."

A friend says that for years she always thought "I look disgusting" when she sat around, and when she took a walk it changed her thoughts to "I look fine, actually." It took her years to realize that the language she linked to the feeling came through years of conditioning to think about her body only in terms of the way that it looked. Decades later she speaks her body's language better and can hear the request to get moving without a side of shame.

Body signals are there, but they are primitive, developed during a time when humans lived very different lives, and are therefore easily misinterpreted. But by paying more attention, you can learn to hear the body's language.

Just as being aware of hanger as a real thing can help us through the negative feelings even when we can't eat right away, being aware of movement's effect on our emotions and mood can be helpful even if we can't change our habits and better meet our needs just yet. Knowing how we work—that movement is part of how we work—can help explain why the world feels the way it does.

This is not to downplay the serious diseases and traumatic experiences some people have. Not all sadness or depression, for example, can be tied to not meeting our movement needs, just as not all anger can be tied to hunger. I'm not trying to position movement as the sole medicine for mental health issues. And there is a compounding factor: getting moving, even when it could make things a little better, can be more difficult because of a mental health issue. Still, we know from our own experiences and a large body of research that the body's daily needs are there, all the time, affecting all parts of our experiences (including serious diseases and trauma), and one of those needs is to get your body moving.

MOVES TO TRY WHEN YOU FEEL FROZEN

If you want to start moving your body but feel frozen by overwhelm, here are a few simple ideas that can get you going:

- Put on fast song and let yourself dance (wiggle those hips and wave those arms) for a few minutes.
- Get up to take out the trash or do a load of laundry. Do any chore that has a lift or heave to it! Bonus if it gets you outside for a few minutes.

- Give yourself a round of applause. Either put music on and clap to the beat or just start your own rhythm; bonus points if you add some foot stamping as well. (Grabbing a couple of pens and drumming on your desk is an easy dose of movement, too.)

- Stand up and take your spine through six simple movements: bend forward while reaching for your toes, arch back, lean to the right and left, and twist to the right and left. Take a deep breath in each shape.

- Walk up and down a flight of stairs or steps in your home or workplace ten times.

- What other ideas do you think should be on this list? Write them down!

If movement affects your appearance in a way that feels positive and this motivates you to move more, so be it. But physical appearance doesn't have to be the reason we focus on movement individually or collectively: we can move for our minds. Moving your body is simply another way to feed your body—and all its parts, *including the invisible ones*—what it needs.

MAKE IT A WANT TO

With so many wants in our lives, our needs have sort of fallen by the wayside. Most of us still eat food and get some sleep each night (even if it's not the amount we'd run best on); the need to move our bodies more is often the ball we drop. Yet until very recently, movement was not a separate "ball" for humans to hold; movement and community were woven into the tasks that made up daily life. If you know you want to move more but are struggling to find time for movement, simply adjust tasks you are already doing to create a dynamic version that increases your physical activity—no juggling required. Give these a try:

- Love time with your partner? Swap out dinner for a date hike.
- If friends (or art or wine!) are your passion, fill a pack with art supplies and a bottle of your favorite vino and trek through a park for a DIY sip-and-paint with pals.
- Meet a friend for coffee—and take it to-go on a garden tour.
- Start a dynamic book club where the conversation is on foot or wheels.
- Take early-rising toddlers on an outdoor sunrise adventure with a portable breakfast.
- Start work meetings with a few team-building exercises that are actually stretches.

- Liven up household chores with a dance party, swinging arms and hips to your favorite tune.

- Host an active potluck, where neighbors of all ages play casual games of Frisbee or basketball and share food.

Experiment together and have fun! Before long, infusing your relationships and daily activities with playful and nourishing movement will become second nature, and this "need to do" will become a "want to do."

WANTING TO WHEN IT'S A HAVE-TO

When I was twelve years old, I would have told you that running a mile was the WORST thing ever. I hated it. But I didn't only hate running *while running*; I began dreading it even when I wasn't doing it. At my middle school we did timed runs every Friday. The first Friday of the month it was one lap; the next, two laps; the next, three; and on the last Friday it would be four laps for our timed mile-long run. It would take me thirteen or fourteen minutes. It hurt me to run. I wanted to barf. I hated being so slow. I hated having to struggle in front of my peers.

When I was in eighth grade, there were two seventh graders who were gifted runners. They could run a mile *in seven minutes*. What was this wizardry? One of them, Nicole (who I recently searched out online—she grew up to be a science teacher and track-and-field coach!) once ran with me and we talked as we ran. This was my first inkling of "when I'm not constantly thinking about how much I hate running, it's so much easier." To my amazement, with her moving alongside me the run not only felt shorter, it was shorter: I shaved *minutes* off my mile run time. *What was this wizardry?*

Although I was a competitive swimmer in my later teens, I continued to *hate* timed runs throughout my entire time in school. Then, when I was about twenty years old at university as a math student, I joined a gym. At the end of the gym-equipment orientation, the dude showed me how to use a treadmill. I had a Walkman music player (true story) so I put on a tape (A TAPE!) and started to run. I started to run by choice and I remember thinking, "I can stop whenever I want, so it's no big deal." I ran a nine-minute mile that day. It blew my mind. I could go faster *and longer* when I chose to run. My categorizing mind just had to move running out of the "I hate this" category. (Since then, I've realized that I dislike "have to" and now work to make sure I'm not conflating the task and the mandate.)

Once I moved it in my mind, I started moving it in my body. I ended up transferring to the biomechanics program in the kinesiology department (for a few reasons, but the idea that I could learn all about the mechanics and physiology of running WHILE running was compelling) and at the end of that year, I ran my first 6:47 minute mile. Over the next ten years I ran (and won some) 5Ks and 10Ks, half marathons and triathlons, and then fifteen years from there I went on to seek more all-day movement, lots of hiking, short- and long-distance walking, and a constant pursuit of keeping generally mobile and strong.

That's how I transformed my personal culture in regards to movement.

Culture is an individual's set of attitudes, values, beliefs, and behaviors, as well as those shared by a group of people, communicated from one generation to the next. We each dwell within an overarching culture but also within smaller subcultures such as a family-of-origin culture, a community or location culture, a religious-community culture, and an education culture, and we all have a personal culture. These cultures are distinct from but also related to each other.

Right now, sedentarism pervades the broader, overarching culture, but subcultures—including our personal culture—can be sedentary, active, or somewhere in between. If a subculture is sedentary it reinforces that aspect of the overarching culture, so what can we do? I'm (obviously) interested in working on sedentarism at the broadest cultural level, but I recognize that the most immediate benefits can be found by changing our personal culture. That's why I address sedentarism at this level as part of my work.

No doubt I'd be moving more if society was set up to encourage or even require us to move more, but I'm happy with my progress to date. My greatest work has been in changing my personal culture, which then has grown into a dynamic family culture. My kids do not have the same sedentary formative years as I did, and here's one example: one day, our family just decided to sign up to run a 10K. We thought the kids would have fun running the first couple of miles and then enjoy walking the rest, but we all ended up running the entire thing. Turns out my ten-year-old, who hasn't enjoyed running in the past, realized that the format of competition and group excitement changed his perception of how hard it would be, and he averaged an eleven-minute-per-mile pace for all 6.2 miles of the 10K. The eight-year-old averaged a fourteen-minute mile for the whole race. (P.S. None of us regularly practice running, but our approach to moving throughout each day seemed to provide us with enough fitness to be able to run when the time came.)

My point isn't running times or even the running itself. We all have our own hurdles to moving more (can you think of a few of your own?). Some hurdles to moving more are physical: disability, lack of practice, undeveloped skill sets. Some hurdles are non-physical and are created by how we think about and perceive movement. Some of those physical hurdles, too, are created by how we think about and perceive movement—as individuals and as a society.

In spite of our personal movement hurdles, a change in perception is powerful, and personal changes can grow beyond ourselves. I now belong to a dynamic family. I worked to grow a personal movement culture and then a dynamic family culture, even though I didn't start in one. That transformation began years before, when a fellow middle schooler took the time to run a mile alongside me and began to change my mind about movement.

When it comes to moving your body, the most important ingredient for success might be your mind. Your mind governs whether you do small exercises that help unkink your neck or your knee so you can take a walk, which, by the way, your mind also decides whether or not to do. Your mind decides whether you join a movement class or reach out to friends to take a weekly walk with you. Your mind decides to organize a community soccer game, sign up for the free guided hike at your nature center, and whether you will actively transport yourself (all or partway) to work or school. Your mind notices that knee is tweaking again, and your mind reminds you of the simple exercises that make all the rest possible.

The shin bone's connected to the thigh bone, sure, but it's also connected to your family, your community, your planet. It is *all connected*.

Each Calf Stretch (page 217) is not just a stretch, it's also a course correction for our individual sedentarism. Every time you choose to move the parts of your body and your body as a whole, you're also shifting, ever so slightly, the collective culture of which your body is a part.

I've mentioned inertia before as part of our problem with sedentarism, because inertia is specifically the tendency for us to continue in the same state, and if we're still, then it's much easier for us to continue being still. But changing our state is possible—we can go from mostly sedentary to mostly active—it just requires effort. The bad news? The effort to change is different from the effort to keep going; it's much harder. The good news? Once you make that effort, your movement keeps you moving. Inertia, our tendency to continue to do as we are doing, can be part of our solution to sedentarism.

Rethinking our positions repositions how we think, which begins a cycle of thinking and repositioning, thinking and repositioning.

What I forgot to tell you at the beginning of the book is that a series of positions is just another way of describing something in motion. And *thinking* about your positions as you make adjustments to them is simply *thoughtful movement*. The more people we invite on this journey with us, the heavier our thoughtful, human-powered collective becomes, which makes it even easier for it to keep going, and easier for even more people to reach for movement and grab on...because they want to.

READ ON

D on't want the book to end? Keep the learning going and read *Movement Matters* next. In this award-winning collection of essays, Katy Bowman goes beyond individual body parts moving well, and explores how each of our bodies is a part of the greater collective of human movement.

If we need movement so much, how does a sedentary collective work? (Hint: outsourced movement.) How does a sedentary culture think about movement and frame scientific questions when it comes to researching movement? If our need to move—our whole body and our parts—is embedded in our biology, how is that need influenced by our other biological needs, for food, shelter, rest, nature, and community?

Most of all, *Movement Matters* offers a new way of thinking about movement, and building movement into our daily lives, not only for our health and the health of our communities and planet, but because it can make the most mundane parts of our lives joyful and something to be celebrated. This book will have you rethinking the way you use your body as a whole each day.

Available in paperback, ebook, or audiobook.

INDEX

pulling a wagon (movement test), 57-58

Q

quadruped/tabletop, 26, 42, 73-74, 97

R

reaching station, 68

rib thrust, (*see* Chapter 2, "Your Ribcage," 29-52*)*, 86, 87, 94-95, 103, 132, 192, 206-207

running, 66, 79-81, 96, 151-153, 224-231, 250, 260, 271-274

S

savasana/lying on your back, 98, 113-120

Shriver, Maria, 259

sleep, 19-22, 114-115, 119, 269

spine (*see* Chapter 3, "Your Spine," 83-120*)*, (*see also* low back)
  - neutral spine, 91-100
  - ranges of motion, 86-87

squatting, 97, 160, 201

stacking movement, 158-159

swallowing, 12-14

swayback (*see* low back: hyperlordosis)

T

tech, (mobile phones, computers, etc.), 25, 59-61, 95, 215

testicles, 168-169

travel, 111, 204-207

TRX strap, 67

V

valgus, 208-210

varus, 208-210

veins, 233-235

W

walking, start 159-160

Women's Alzheimer's movement, 259

REFERENCES

To save paper and because you'll be accessing these sources online anyway, we've put chapter-by-chapter references and links for *Rethink Your Position* online at nutritiousmovement.com/RYP-references.

IMAGE CREDITS

Images are cited by page number. All uncredited images were provided by the author.

Chris Eckert: 36

Mahina Hawley: 57, 59, 60(2), 61(2), 85, 102, 124, 136*, 137*, 138*, 139*, 222

Intarapong/Shutterstock: 209

Jen Jurgensen: 45, 46, 92, 105, 118, 141, 171, 173, 174, 175, 218(2), 223

Michael Kaffel: 256

Mark Kuroda: 148

Cecilia Ortiz: 10, 13, 34, 42, 43, 44, 108, 111, 128, 129, 130, 133, 135, 183, 195, 218(1), 219, 239, 240, 241, 243

Andrey Popov: 31

SrdjanPav/iStock: 193

*these four images were composites, with art by Mahina Hawley and graphics by ©decade3d/123RF.COM

ABOUT THE AUTHOR

Bestselling author, speaker, and a leader of the Movement movement, biomechanist Katy Bowman, M.S., is changing the way we move and think about our need for movement. Her ten books, including the groundbreaking *Move Your DNA*, have been translated into more than sixteen languages worldwide.

Bowman teaches movement globally and speaks about sedentarism and movement ecology to academic and scientific audiences such as the Ancestral Health Summit and the Institute for Human and Machine Cognition. Her work is regularly featured in diverse media such as the

Today Show, CBC Radio One, the *Seattle Times*, NPR, the Joe Rogan Experience, and *Good Housekeeping*.

One of Maria Shriver's "Architects of Change" and an America Walks "Woman of the Walking Movement," Bowman consults on educational and living space design to encourage movement-rich habitats. She has worked with companies like Patagonia, Nike, and Google as well as a wide range of non-profits and other communities to create greater access to her "move more, move more body parts, move more for what you need" message.

Her movement education company, Nutritious Movement, is based in Washington State, where she lives with her family.